5

M
Al
P.C
Wi
Tel/Fex (250) 766-4718

METHODS OF
PLACER
MINING

Garnet Basque

SUNFIRE PUBLICATIONS LIMITED
Langley, B.C.

SUNFIRE PUBLICATIONS LIMITED
P.O. Box 3399,
Langley, B.C.   V3A 4R7

PRINTING HISTORY
First Printing — August, 1983
Revised Second Printing — June 1994
Printed in Hong Kong

PHOTO CREDITS
B.C. Government: Front cover, main photo.
Sunfire Archives: 8, 13, 15, 21, 24, 26, 28, 29 & 37.
Yukon Archives: 32 & 42
California Div. of Mines & Geology: 39.
B.C. Dept. of Mines & Petroleum Resources: 45, 46, 49 & 50
Prov. Archives of Alberta: 55.
National Archives: 57 & 83.
Keene Engineering: 61, 65, 69 (top), 70, 72, 76 (top), 77 & 78.
Placer Equimpent Mfg.: 69 (bottom), 73 & 76 (bottom)
E.F. Domine: 79.

DRAWINGS
All maps and drawings used in this book were drawn by the author, with the following exceptions:
   FIG. 2 THE ROCKER, p. 44, *Notes on Placer Mining, Bulletin No. 21.*
   FIG. 5 THE SLUICE, p. 51, *Notes on Placer Mining, Bulletin No. 21.*

OTHER BOOKS WRITTEN BY GARNET BASQUE
Lost Bonanzas of Western Canada Vol. 2
Ghost Towns & Mining Camps of the Boundary Country
West Kootenay: The Pioneer Years
Fraser River & Bridge River Valley
British Columbia Ghost Town Atlas
Yukon Ghost Town Atlas
Gold Panner's Manual
Canadian Treasure Trove

CANADIAN CATALOGUING IN PUBLICATION DATA
   Basque, Garnet
      Methods of placer mining

      Includes bibliographical references
      ISBN 0-919531-40-7

      1. Hydraulic mining—Handbooks, manuals, etc.
      2. Hydraulic mining—British Columbia--Handbooks, manuals, etc.
      3. Gold mines and mining—History—Handbooks, manuals, etc. I. Title.
   TN421.B38 1994   622'.3422   C94-910428-0

CONTENTS

INTRODUCTION

"GOLD has been called the 'pioneer' of civilization, because it led men into the wilderness, across unknown seas, and over wide prairies and desert plains. It has been pointed out that every 'gold rush' was the beginning of new states and countries; that the discovery of gold in some far-off locality did more to populate the land than the most energetic efforts of immigration agents or government officials."

These comments, published in 1897 in *Klondike—The Chicago Book For Gold Seekers*, reveals an important, though often overlooked clue in the modern prospector's search for gold. When the price of gold soared to over $800 U.S. an ounce early in 1980, thousands of would-be prospectors and gold panners got the "itch" to dig a little of the precious metal. Unfortunately, most simply had no idea of what to look for or where to begin, resulting in many unrewarded and wasted hours being expended on a helter-skelter search for gold-bearing gravels. Yet, as pointed out in the opening paragraph, the answer should be embarrassingly obvious. Hundreds of thousands of prospectors flooded California in 1849; Australia in 1851; the Fraser River in 1858; Arizona in 1874 and the Klondike in 1897. What attracted these people like the proverbial flies around a molasses barrel? Gold—of course!

Since it is extremely unlikely that a major discovery of gold will ever again fire the imagination of men, we have no option but to re-work the areas of historic discoveries. Before you can become a successful gold panner or prospector, therefore, you must first become a historian. You must know where gold has previously been discovered, since your chances of finding gold in the same region is immensely more probable than a haphazard search in a barren area. Thus, we come to the purpose of the first chapter. In the next few pages I will relate a brief history of the world's major gold rushes and gold-bearing areas. This is by no means the definitive work on the subject, nor is it meant to be. It would take a book many times this size to cover the subject in detail. Rather, Part One is meant as a starting point for your research. Many books have been written on local or world gold rushes from which you should be able to extract some valuable clues. In the meantime sit back, relax, and see what happened to the people and places where gold rushes have occurred.

PART I
GOLD RUSHES
OF THE WORLD

ENTURIES upon centuries have come and gone since the stories of fabulous gold discoveries first fired the hearts and imagination of men. Though transportation in ancient times was extremely difficult, restricting access to the diggings, gold rushes still occurred. It is difficult to imagine that the vast quantities of gold which existed in Judea, Babylon, India, Persia and Egypt were merely the results of gradual accumulation. It is far more likely that the bulk must have been the results of rich gold discoveries.

The bible, for example, tells us about the extravagant use of gold by Solomon. When he first came to the thrown he virtually plastered the temple with gold. These "ziggurats" were built in the valleys of the Euphrates and Tigris rivers, and at Babylon. Of Sumerian origin, the largest was some 700 feet high and, according to *Gold—From Caveman To Cosmonaut,* "was built in seven tiers, the topmost crowned with solid gold and reserved for the personal occupancy of the Sun God. The lower levels contained much furniture and many artifacts of gold. Reliable estimates indicate that the Babylon ziggurat contained at least 50 tons of gold." At $500 an ounce, this ziggurat alone would be worth $600,000,000 today. When one considers the other ziggurats, the idols and vessels of gold, it is unlikely this wealth was accumulated without some massive discoveries somewhere. Considering the undeveloped and primitive methods of mining then in use, it is also certain that these discoveries were alluvial or placer deposits, as future gold rushes were. Where these discoveries occurred, however, is unknown today, although Ophir, Punt and the River Pactolus are often touted as the source.

Ophir is the first El Dorado of which we have any record, but

Solomon's temple in Jerusalem.

it is sketchy at best. Except for the information that it was a gold-producing country, there is no data by which it can be even approximately located. From *Gold—From Caveman To Cosmonaut,* we learn that "In spite of years of research by archaeologists and other scientists, it is still uncertain where Ophir was. It is variously thought of as being in Asia Minor, India, Ceylon, or the Malay Peninsula; in southern Egypt, Ethiopia, or the eastern coast of Africa. Some even hold the opinion that it was in Peru."

Zimbabwe, near the Zambesi in Rhodesia, was for a time believed to be the site. However, archaeologists are now convinced that the gold mining activity of the area is too recent to be credited to either Ophir or Punt. The location of the River Pactolus has likewise been lost in time. The Pactolus River gold was alloyed with a considerable amount of silver, resulting in a whitish metal called electrum.

The first rush of gold seekers of which we have authentic historical record took place from Spain to the countries discovered by Columbus. Although no gold mines existed on the islands or those portions of the continents visited by Columbus, the natives he encountered wore ornaments of gold. Cortez gathered a great amount of this, obtained mostly from South America, although neither he nor his men undertook to mine.

Spain's golden wealth began to emerge when Francisco Pizarro, illiterate and hungry for gold, set sail for the New World. In 1532 Pizarro landed in Peru with less than 200 men. He was greeted by Atahualpa, the Inca chief and 10,000 warriors. As the Spaniards neared the chief they attacked, instantly killing most of the native leaders, and holding Atahualpa for ransom. For his freedom, Atahualpa offered to have the large room in which he was held prisoner, filled with gold, and two smaller ones filled with silver. Pizarro agreed to these terms, amassing a fortune in gold and silver, then treacherously had him executed. To divide the treasure among his men, Pizarro found it necessary to melt the beautiful goldwork into ingots. In one of the worst cultural tragedies in history, the Inca craftsmen were forced to systematically melt the superb ornaments on which they had so lovingly laboured. Such was the quantity of gold amassed that the task is said to have taken three months to complete. From their plunder, the Spanish army was able to send home four times as much gold as the followers of Cortez had from Mexico. But none of this great wealth was obtained directly from the mines.

Numerous expeditions were undertaken during the first century of the New World's history for the express purpose of finding gold. Despite this, gold was not discovered anywhere within the present territorial boundaries of the United States, in fact nowhere north of the Rio Grande, until 300 years after the death of Columbus. But the lure of gold did attract hundreds of adventurers from across the Atlantic. They braved the dangers of the unknown wilderness in their futile attempts to locate the fabled El Dorado of which the Indians spoke. Gradually the hope of discovery vanished, and for 200 years the golden phantom did not reappear. It was not until the 19th century that North America began to reveal a glimpse of the fabulous wealth it contained. For the most part, these discoveries were the results of accidents— sometimes obscenely so.

❀   ❀   ❀

NORTH CAROLINA
The North Carolina State mint report of 1793 provides us with the first information of United States gold production, but it reveals no clues as to the source of the gold. The first recorded discovery of gold, however, is well documented. It began innocently on a sunny afternoon in 1799, as 12-year-old Conrad Reed, accompanied by a sister and younger brother, set out for Meadow Creek, a small stream near his father's farm in Cabarrus County. Reaching the stream, young Conrad noticed a yellow stone glistening in the

water. Unaware of its nature or value, but intrigued by its colour, the youngster carried it home. The boy's father, John, was unable to identify it, so he took it to a silversmith in Concord for appraisal. Apparently the silversmith's credentials were somewhat questionable, for he was also unable to determine the metal's value. So the 17-pound nugget was returned to the Reed farm, and for three years it was used as a door stop.

In 1802 the old farmer had occasion to visit Fayettville and carried the nugget with him. A jeweller promptly recognized it as gold and offered to have it refined for Reed. When Reed next visited Fayettville the jeweller displayed a large bar of gold which he was interested in buying. Reed, unaware of its actual value, but determined to make a nice profit, demanded what he considered a "big price"—$3.50. The jeweller must have had difficulty remaining calm as he paid for his bargain.

Upon returning home, Reed hurried to the creek where the original nugget had been discovered and found gold all along the stream. Forming an association with three German neighbours, they began to mine the creek. In 1803 they uncovered a piece of gold that weighed 28 pounds; numerous nuggets of various other sizes were also found. The whole surface of the ground along the creek's bank for nearly a mile was rich in gold. By 1825 mining was in full swing, and from 1829-1855, the most productive gold-producing years, 393,119 ounces were mined. The early production was from placers, but in 1831 quartz veins were discovered which yielded large quantities of gold. By 1850 several important lode mines were in operation; Reed, Gold Hill, Kings Mountain, Rudisit, Conrad Hill and Phoenix. Although closed during the Civil War, most were reactivated after the cessation of hostilities.

North Carolina's golden deposits were traced southward as far as the borders of the Cherokee Reservation in Northern Georgia. In Rowan County, Gold Hill mining operations were commenced in September 1842, where some very rich veins were opened. From January 1843, to July 1951, gold valued at $801,665 was mined there. It was the Cherokee Reservation, however, which held the richest deposits, drawing hundreds of prospectors into the Indian's domain. This encroachment was bitterly protested, and Georgia dispatched a large police force to expel the invaders. But it proved useless. Reckless, dissipated men from all quarters of the country flocked in, prowled about the woods, set up log huts and shanty groceries on all the streams, and even the Federal troops were powerless to keep the lawless hoards west of the Chestatee. Finding that no protection of the Indians by police measures was feasible, in 1830 the State adopted the Indians, reservation and all, and constituted the region a county. The State then divided the mineral lands into 40-acre-lots and sold them by lottery. However, here as elsewhere, it was soon realized that the gold could not easily be obtained without hard work. The worthless, lazy and dissolute majority of the early horde of invaders gradually drifted away, while only the small minority of newcomers remained. The population, like the dirt, was slowly panned out, and the current of events carried the dross away.

North Carolina has produced more gold than any other southeastern state. The numerous deposits were located within two physiographic provinces—Piedmont and Blue Ridge. The counties of Mecklenburg, Rowan, Cabarrus and Davidson, in Piedmont province contained the most productive mines and richest deposits. Burke and Translyvania counties contained the primary deposits in Blue Ridge province. These gold-bearing areas consist mostly of granite, mica schist, gneiss and hornblende gneiss.

❀   ❀   ❀

GEORGIA

Georgia is one of the rare locations on the Atlantic side of the continent where gold has been found in any quantity. The first discovery in 1829 was in the Nacoochee River valley in White County, and near Dahlonega in Lumpkin County. The early pioneers had no ready market for the gold they panned but, lacking lead, they moulded bullets out of the precious metal.

In 1831 a man named Wilpero, observing the similarities of the surface, foliage and streams to the gold-producing region of

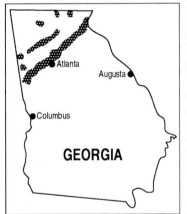

North Carolina, began to search for gold. He was rewarded with a find of considerable quantity in Habersham County. From *Principal Gold-Producing Districts Of The United States* we learn that "Thousands of people flocked to these areas to wash gold from the superficial deposits along creeks and streams. By 1838, production was large enough for the government to establish a branch of the mint at Dahlonega." Gold mining, halted by the Civil War, concentrated on lode mining after hostilities ended. From 1916-1934 the annual production was only a few hundred ounces.

Georgia's gold-bearing regions occur in several mineralized zones in the crystalline schists and gneisses that underlie the northern third of the state. There were approximately 500 sites purported to contain gold in this region, but only three localities—the Creighton mine in Cherokee County, the Dahlonega district in Lumpkin County, and the Nacoochee district in White County—have had significant production."

VIRGINIA

The earliest mention of gold in Virginia was made by Thomas Jefferson. In 1782 he recorded the discovery of a piece of gold weighing 17 pennyweights that had been found along the Rappahannock River in southern Piedmont. The Virginia Mining Co. was the first gold-mining company incorporated in the state. Owned by a New York firm, they operated the Grasty tract in Orange County from 1831 to 1834. According to mint records, $2,500 in gold was mined in Goochland County in 1829. In the early 1830s, other deposits were discovered along the gold belt. Unlike other southern states, gold production did not regain its former level of productivity after the Civil War.

"The gold-bearing areas of Virginia," according to *Principal Gold-Producing Districts Of The United States*, "are east of the Blue Ridge Mountains in a belt 15 to 25 miles wide and 200 miles long. The ore deposits are in northern-trending schists and gneisses of sedimentary and igneous origin. They are cut by auriferous quartz-pyrite veins that in general are concordant with the strike and dip of the host rocks." The most important deposits are contained in the counties of Fauquier, Fluvanna, Goochland, Orange and Spotsylvania.

❀   ❀   ❀

ALABAMA

The gold discoveries in Georgia stimulated interest in prospecting the similar-appearing crystalline rocks of Alabama, and by about 1830 the first discoveries were made. In the 1830s and 1840s, thousands of people were working the deposits at Arabacoochee and Goldville, but this boom collapsed when the California placer discoveries lured away most of the miners."

Gold mining accelerated in 1874, when copper was discovered in large quantities; in 1904, when the cyanide process was introduced at Hog Mountain; and again in 1934, because of the increased value of gold.

❀   ❀   ❀

CALIFORNIA

The discovery of gold in California is generally credited to James W. Marshall, whose highly publicized placer discovery ignited the mad stampede of 1849. But gold was not new to California. For three centuries there had been rumours about fabulous mineral wealth in the Sierras. Mexicans had mined small amounts of placer gold from the Colorado River, and lode gold from the Cargo Muchacho Mountains, as early as 1775. In the 1820s-1830s, small nuggets were often obtained from the Indians. In the 1830s, gold was discovered in Los Angelas County. Unfortunately, these activities had all occurred in sparsely-populated, remote regions

METHODS OF PLACER MINING
12

(Above) Sutter's Mill, the scene of one of the greatest
gold discoveries in history. (Inset) John Sutter.

that were still under Mexican control. The remoteness of the discoveries, combined with a lack of direct communication, prevented the information from spreading. In 1848, all that changed.

James Marshall, building a sawmill in partnership with John Sutter, would make the discovery that would electrify the world. "On the morning of Monday, January 24, 1848, Marshall was walking in the tail-race, when the rush of water was carrying away the loose dirt and gravel, and saw on its rotten granite bedrock some yellow particles, and picked up several of them. The largest was about the size of grains of wheat. They were smooth, bright, and in colour much like brass."

On the evening of February 2, 1848, Marshall rode into Sutter's fort on the Sacramento River, his mud-splattered horse foaming at the mouth from being driven hard. Taking Sutter aside, Marshall showed him about half a thimbleful of yellow grains of metal.

Although Sutter had received a grant of 50,000 acres by the Mexican government, the site Marshall had chosen for the mill, some 45 miles from the fort, was not on his property. Sutter wanted desperately to keep the find a secret until he could get in his harvest. But Marshall's cook leaked the information and the harvest was never gathered.

At first there was little excitement, probably because the news was not easily communicated. But on May 15, Sam Brannan rushed through San Francisco waving a bottle of dust, and by mid-June most the town's men had set out for the mines. Sutter's worst fears were quickly realized. His oxen, hogs and sheep were stolen by hungry men and devoured. No hands could be procured to run the mill, and his lands were squatted on and dug over. Sutter wasted his remaining substance in fruitless litigations trying to recover them. To carry on the legal warfare, he was compelled to sacrifice or mortgage the parts of his estate not seized by the gold diggers, until, little by little, his magnificent property melted away, leaving him all but destitute. Sutter died in Washington, D.C. on June 17, 1880, a poor man.

Marshall did not fare much better. His property was seized and divided into town lots. Marshall's discovery, which produced $65,000,000 in one year alone, and averaged $25,600,000 a year for 17 years, brought him neither fame nor fortune.

The earliest gold-seekers came from Mexico, the South America coast and the Sandwich Islands (Hawaii). By the winter of 1848-49, gold from the diggings began pouring into Valparaiso, Panama and New York, carrying the excitement east. In the spring a flood of adventurers made their way to California across the plains or the Isthmus of Panama. It is estimated that 100,000 men reached California during 1849, adding 12-fold to the population and 50-fold to the productive capacity of the territory.

By January, 1849, 90 vessels, carrying 8,000 passengers, had sailed from various ports, bound for San Francisco, and 70 more were advertised to sail. Before the year ended, 549 vessels entered the Golden Gate. Early in 1849 the population of San Francisco swelled from 2,000 to 14,000. Four hundred sailing vessels were abandoned by their crews at their anchorage in the bay. Labour was $10 a day. The yield of the mine in 1849 was probably no less than $18,000,000. The excitement ran so high that, between July 1848 and the end of 1848, 716 out of 1,290 soldiers in northern California deserted for the diggings. They had been receiving a paltry salary of only six dollars a month, so it is hardly surprising that most decided to try their luck in the goldfields.

The emigration to California continued unabated for three years, before falling off in 1854. By the end of 1850 some 50,000 men were engaged in gold mining. From 1852-53 the number

(Above) The Mexican arrastra was grinding ore long before gold was discovered in California.
(Below) This modern version of the arrastra was being used in the Osoyoos mining district of British Columbia in 1923.

increased to 100,000, and the annual production of gold reached $65,000,000.

These discoveries, and the resultant rush of population to California, initiated some wild and lively times. Lots in San Francisco were said to be worth gold coin enough to carpet them. Speculation ran wild. All forms of gambling were recognized as legitimate business, while adventurers and criminals flocked in. Society became chaotic, and at length self-preservation required the organization of the celebrated vigilance committee to enforce order.

The first and richest placers, on the American, Yuba, Feather, Stanislaus and other small streams in the heart of the goldfields, yielded as much as $1,000-$5,000 a day per man. However, these surface operations were relatively small, and when they were quickly depleted, attention was then diverted to the large-scale gold-bearing deposits of the Tertiary age. By 1852 large-scale hydraulic operations began, and for many years this was the chief gold-production source of California. The total gold production from 1848 to 1856 is estimated at $450,000,000, calculated at $20.67 an ounce. By 1870, largely because of mechanical mining, the total had risen to $1,000,000,000.

As the gold-bearing gravels were being depleted, prospectors began searching for the primary source of the deposits. Gradually a massive mineral belt was identified which became known as the Mother Lode. From *Klondike—The Chicago Book For Gold Seekers,* we learn that this area ". . .covers an area about a mile wide and about 120 miles long extending from Mariposa County due southeast of Sacramento to northern El Dorado County, approximately on a line from Sacramento to Lake Tahoe, and passing a few miles north of Columa," where Sutter's mill had been located. Lode mining gained prominence in the 1860s, and from 1884-1918 gold-quartz veins were California's chief source of gold.

The western slope of the Sierra Nevada Mountains, in central California, was the most productive gold-bearing area. "This region contains the Tertiary channel gravels, the Quaterary stream deposits, the Grass Valley-Nevada City lode district, the Alleghany and Downieville districts, and the complex vein system of the Mother Lode, East Belt and West Belt.

"The auriferous gravels of California are of two general types: buried placers of Tertiary age and normal stream placers of Quaterary age. The gold was derived from many gold-bearing veins, including those of the Mother Lode." An extended period of erosion followed during Tertiary time which nearly levelled the mountains. The golden particles from the eroded veins were then concentrated in stream channels. California offers excellent potential for the gold panner or small placer operator, as significant quantities of gold have been mined in 41 of the 58 counties. In 1933, for example, there were 993 placer mines in operation, as well as 797 producing lode mines.

❀   ❀   ❀

NEW MEXICO
Mining, of one sort or another, was being carried on in New Mexico long before discoveries were made in any other western state.

Placer mining commenced in the Ortiz Mountains, south of Santa Fe, as early as 1828. New placer discoveries were located along the foot of the San Pedro Mountains in 1839. The first recorded lode mining operation occurred in the Oritz Mountains as early as 1833. At the end of the Mexican War in 1846, New Mexico became a part of the United States. But, "because of its isolation, the general lack of knowledge of the region, and the hostility of the Apache Indians, it was not until about 1860 that prospectors and miners were attracted to the region."

Despite treacherous Indian raids, which frequently interrupt-

ed mining, several million dollars worth of placer gold is said to have been mined prior to 1848. From 1848 to 1965 New Mexico's gold production was about 2,267,000 ounces. New discoveries in the middle and late 1860s and in the 1870s stimulated mining. "In 1865 placers, and, soon afterward, quartz lodes were found in the White Mountains in Lincoln County; in 1866 placer deposits were discovered at Elizabethtown in Colfax County. In 1877 placer and gold-quartz veins were found in Hillsboro."

New Mexico's mineral belt is of variable width and extends diagonally across the state. The gold producing districts are distributed from Hidalgo County in the southwest corner to Colfax County along the north-central border. This mountainous terrain is a zone of crustal disturbance in which "the rocks were folded and faulted and intruded by stocks, and laccoliths of monzonitic rock. The gold deposits were probably derived from the weathering of these deposits."

In New Mexico, 17 districts in 13 counties yielded more than 10,000 ounces of gold each through 1957.

       ❀   ❀   ❀

NEVADA

After the California gold rush, many down-and-out prospectors returning east happened to camp near Washoe, then part of Utah. Among them were two brothers, Hosea and Allen Grosch, who arrived in 1853. They struck silver in 1856, but died a year later before they could begin mining.

In January 1859, four prospectors, working at the head of Gold Canyon, discovered pay dirt. A modest gold rush began, although gold production has been greatly exceeded by silver production.

Some 4,000 people arrived at the Nevada diggings that summer and autumn, and Virginia City was born. "From 1859-80, the Comstock produced $300,000,000 in silver and gold. As for the site of the Big Bonanza, its richest ore was worth $1,000 to $10,000 per ton of ore, and in all the Consolidated Virginia and California made $150,000,000 and paid out $78,148,000 in dividends over 22 years." Except for the Comstock, mining declined steadily after 1880 until the Tonopah silver deposits were discovered in 1900.

The major gold districts are Goldfield, Silver Peak, Aurora, Rawhide, Jarbidge, National, Round Mountain, Manhattan, Delemar, Wilson, Petosi and Lynn.

❦ ❦ ❦

AUSTRALIA

Although it is highly unlikely that the readers of this book will search for gold in Australia, I have decided to include a brief history of its gold discoveries for two reasons. First, it was a gold rush of major importance and second, it points out the undeniable benefits of being a student of history; for this discovery, unlike many others, was not the result of accident.

E.H. Hargraves had been one of the thousands of adventurers who swarmed into California in 1949. Although he had been unsuccessful in the goldfields, he was amazed by the striking similarity between the geographical features of California and those of his native Australia. Sir Roderick Murchison had predicted that gold would be found in Australia and, in 1850, Hargraves returned to search for it in an area he had known before going to America.

The promising area lay along Lewis Pond Creek, a tributary of the Macquarie River. With the assistance of a young man named Johnnie Lister, he prospected along the creek. At first their labours were unrewarded, but when Hargraves had to return to Sydney, he showed Lister and James Cook (who had joined them on the second day), how to construct a rocker, which Hargraves had seen in operation in California. With the assistance of this new tool their persistence finally paid off, and on April 7 they successfully washed out $75 worth of gold. A geological engineer quickly confirmed the discovery and the Australian gold rush was on.

Within a week $10,000 in gold was washed from the stream, including one nugget weighing over 46 ounces, and in just over a month a thousand men were working Lewis Pond Creek. Immigrants by the thousands began arriving from other countries, reversing the mass exodus to California two years earlier. By 1852, 45,000 pounds of gold was produced in New South Wales, and the population had doubled. Simultaneously, all articles of commerce advanced; wheat quadrupled in value; potatoes rose from seven shillings to 21 shillings a hundred weight; and freight from Sydney to the mines increased from $12 to $150 a ton.

Melbourne soon became jealous of Sydney and a "Gold Discovery Committee" offered a generous reward for the discovery of a goldfield within the province of Victoria. It brought spectacular results. In August 1851, James Edmonds, who had also gained his experience in the goldfields of California, began searching gold formations. Ninety miles north of Melbourne he panned a few dollars. This information encouraged others, and a short time later two prospectors discovered the rich deposits of Ballarat. Ten thousand prospectors flocked to the site, which maintained the reputation as the greatest gold camp in the world until surpassed by Mount Alexander and Bendigo Creek. Melbourne, which had a population of only 4,500 in 1841, grew to 39,000 by 1852, and exploded to 139,000 by 1860. By 1855, 1,250,000 immigrants had arrived in Australia. They faced greatly inflated prices: flour, worth $100 a ton at the dock, demanded $1,000 a ton at the mines; oats rose eight-fold; and mining tools brought whatever the sellers demanded.

The finder of one of the richest veins was a man who had been prospecting in the bush unsuccessfully for a long time, and was returning to Perth disenchanted. One night as he was camped in the wilderness, his horse grew restless. As he attempted to quiet the animal, he tripped over what he first mistook for a large stone. On closer observation, however, he realized it was an enormous nugget of almost pure gold. Within a month six men, using the roughest tools, had extracted $250,000 in gold.

The yield from the Australian goldfields swelled month by month and year by year until, in 1856, the export from Melbourne alone was over $65,000,000. In the same year the mint at Sydney received $7,500,000 in gold from the mines and New Zealand produced $10,000,000.

On June 15, 1858, a party of 24 men, working at the Bakery Hill mine near Ballarat, discovered a huge gold nugget at a depth of 180 feet. The "Welcome" nugget, as it was named, measured 20"x12"x7" and contained 185 pounds of almost pure gold. It was exhibited in Melbourne and London for some time before it was finally melted down to yield 2,159 ounces of gold.

This nugget remained the largest ever discovered until February 9, 1869, when a 190-pound nugget, christened the "Welcome Stranger," was unearthed only inches below the surface in the

Moliagirl goldfield near Victoria. Discovered by John Deacon and Richard Oates, it assayed 98.66% pure and yielded 2,268 ounces of gold.

The "Holtermann" nugget, weighing more than 200 pounds, and an unnamed boulder of gold weighing 630 pounds, were discovered in New South Wales around 1872.

Other large Australian nuggets include the "Pleasure," weighing nearly 135 pounds; the "Viscount Canterberry," weighing 92 pounds; and the "Kum Tow." Over the years the Australian goldfields continued to produce hundreds of nuggets over a pound, dozens over six pounds, and dozens more over 30 pounds each.

Large nuggets of unknown origin are still periodically discovered in the Victoria goldfield. In the 1950s, for example, a $2,000 nugget was found protruding through the pavement in a main street. Another man found a $10,000 nugget while digging in his backyard.

The gold in Australia was found in silurian rocks, especially in the more ancient beds, and in the gravel and debris of these rocks. It also occurred in the gravel of the drifts of the miocene of the territory. The greatest finds were made in quartz veins towering a lower silurian schist rock formation, on spurs of the great Australian Cordillera.

❀ ❀ ❀

MONTANA

The first gold strikes in Montana were made in 1852 in gravels along Gold Creek in Powell County. Henry Thomas sank the first shaft there in 1860, but it was not until two years later, when rich placers were discovered along Grasshopper Creek, near Bannack, that the gold rush began in earnest. Other discoveries, both placer and lode, came in rapid succession. "In May 1863, the very rich deposits along Alder Gulch near Virginia City in Madison County were discovered; these proved to be the most extensive and the most productive placers in Montana. The Last Chance (Gulch) placers on the present site of Helena in Lewis and Clarke 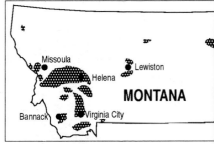 Country, among Montana's most productive placers, were discovered in the summer of 1864 as were the placers in the Butte district in Silver Bow County."

These rich discoveries stimulated activity in the area, and placer mining flourished from 1862 to 1876, producing over

$150,000,000 in gold. Grasshopper Creek was the scene of the first dredging operation in 1895. Dredges operated by electric power were in use from 1906 to the Second World War. Since then, until recently, very little placer mining had been done.

In Montana, 54 mining districts in 17 counties have each produced more than 10,000 ounces of gold. Four districts—Butte, Helena, Marysville and Virginia—have produced more than 1,000,000 ounces, with 27 other districts producing between 100,000 and 1,000,000 ounces.

❀   ❀   ❀

COLORADO

The first indication that gold existed in Colorado was mentioned in a secret report delivered to Zebulon Pike in 1807, while he was being held prisoner in Santa Fe. From 1849 to 1857 the Cherokee Indians were reputed to have obtained gold from the headwaters of the South Platte River. In 1850 a party of 120 discovered gold near the future site of Denver. But they were headed for the California gold rush and decided to continue on.

In the spring of 1858, the Russell brothers, returning from California, led a party of prospectors along Cherry and Ralston creeks and along the South Platte River near the present site of Denver. Although only a small quantity of gold was found, news of the discovery spread and a new gold rush was on.

There was no justification for a major rush, but through the

A placer gold mining operation in Russell Gulch, Colorado, in 1877.

METHODS OF PLACER MINING

21

glorified fabrications of the local Chambers of Commerce, it was impossible not to be impressed. The region was in a severe recession, and the towns in the area concluded that a gold rush might just be the stimulant they needed. Through hired agents they falsely publicized the richness of the discoveries. "Gold exists throughout all this region. It can be found anywhere. . .In fact, there is no end of the precious metal. Nature herself would seem to have turned into a most successful alchemist in converting the very sands of the streams to gold."

Spurred by these official reports, thousands swarmed into the area, and by Christmas of 1858 several settlements has been founded in the Denver area. By the spring of 1859 some 100,000 people had moved in, but half would be gone again by the end of the year.

In January 1859, commercial quantities of gold were discovered by George A. Jackson near the mouth of Chicago Creek. On May 6, John Hamilton Gregory found outcrops of veins with residual deposits of gold in the drainage basin of North Clear Creek near Blackhawk, and in early June, Green Russell made a new discovery of placer gold in Russell Gulch near Central City. Soon 5,000 men and a few hundred sluices were 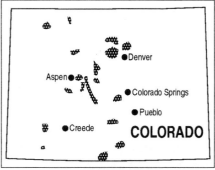 at work near Gregory Gulch. Production during 1858-67 reached $25,000,000, $10,000,000 of which was from lode mining. From 1864-68 the Colorado goldfields experienced a decline, and the state began to lose its immigrants.

Some scattered to other parts of the state and made new discoveries. A stampede to the San Juan Mountains, in southern Colorado, was initiated when gold was discovered there in 1870. "In 1875 major ore discoveries were made at Lake City, Ouray, and Telluride. In the middle and late 1870s rich ore deposits were discovered on the east side of the Sawatch Range in the Monarch and Chalk Creek districts, in the Rosita Hills near Custer County, in the Kokomo and Breckenridge districts in the Tenmile Range, and at Aspen on the west side of the Sawatch Range."

In 1891 Colorado's economic difficulties received another much needed boost when new deposits were discovered at Cripple Creek. Over the wind-swept mountain tops, or waist-deep through snow in the gulches, the adventurous treasure-hunters poured into the district. For many years this was one of the world's leading gold-mining regions, producing more than $400,000,000 since its discovery.

❀   ❀   ❀

IDAHO

Gold was first discovered in this state along the Pend O'Reille River in 1852. However, it was the discovery by a man named Pierce, in the Idaho Panhandle in 1860, that is recognized as the most significant. New deposits were quickly located in Elk City, Orofino, Warren and the Boise Basin the following year.

Although none of these discoveries were sufficient to create a "gold rush" in the true sense, thousands arrived from California, Oregon and British Columbia. As in other areas, the attraction of the precious metal was a major factor in the settlement of the region, but the camps were so remote, they drew some of the wildest and unsavoury characters.

The 1833 report of the Director of the Mint stated that 240 operations were active in the state, but the location of most of these is unknown. Yet, although gold is widely distributed in the state, the deposits are not rich. Despite this, however, placers were the major source of gold in Idaho before 1900. While most of these were exhausted during the "rush" years of 1860 to 1870, geological authorities still believe a great deal of gold remains to be found. "Abandoned mines, tailing dumps and unexplored areas are potential sources of much gold. The State Bureau of Mines at Moscow, Idaho, has published numerous excellent reports on the subject, including descriptions of many old locations which were abandoned for one or another reason but are definitely known not to have been exhausted at the time of shut down."

❀   ❀   ❀

WASHINGTON

While searching for a railroad route through the Cascade Mountains in 1853, Capt. George B. McClellan's survey discovered gold in the Yakima Valley. A mini-rush developed in the Colville area in 1855, and numerous placer strikes were reported by the Portland *Oregonian*. There were numerous other deposits located throughout the 1880s, but these had been mostly worked out and abandoned by 1900.

In 1871 the base of Mount Chopaka was the site of the state's first lode discovery. Numerous other lodes were located from 1871 to 1898, and the mineral from these sources produced substantial quantities of gold.

"An unusual feature of Washington's gold is the unique crystalline wire gold that is seen as nuggets from the Liberty district in the central part of the state. These nuggets occur in degrees of roundness, varying from well-rounded to very rough.

The finest specimens of crystalline gold are mined from lode deposits in the district."

For the Washington prospector, it is worthy to note that placer gold has been discovered in 23 of the state's 39 counties; lode gold in 25.

❀   ❀   ❀

BRITISH COLUMBIA

Although gold had been discovered as early as 1850, it was not until 1858, when the steamer *Otter* sailed into San Francisco with 800 ounces of gold, that the Fraser River gold rush began in

The Fraser Canyon eight miles west of North Bend, B.C. There were 94 bars, flats and riffles on the Fraser between Sumas and a point about 30 miles above Lillooet.

METHODS OF PLACER MINING

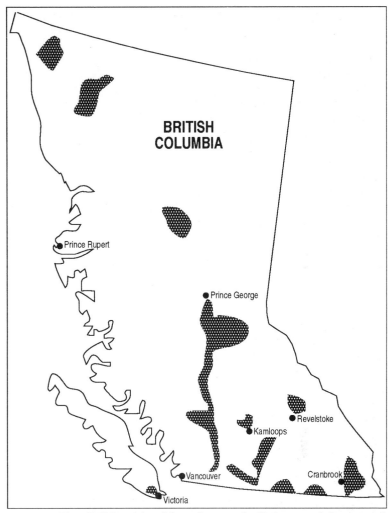

BRITISH
COLUMBIA

Prince Rupert

Prince George

Revelstoke

Kamloops

Vancouver

Cranbrook

Victoria

earnest. The California gold rush was declining, and frustrated prospectors swarmed north. On April 25, 1858 the paddlewheel steamer *Commodore* docked at Fort Victoria with 450 men on their way to the Fraser River. She was followed by numerous others, and in May, June and July an estimated 31,000 people arrived in the new colony. Every bar along the Fraser's course for 140 miles was being worked. Two flourishing towns, Yale and Hope, were established along the river banks. By November 19, 1858, the mainland Colony of British Columbia had been created, prudent law enforcement agencies were quickly instituted, and resolute steps taken to open routes of communication, pacify the Indians,

The Cameron claim, Williams Creek, British Columbia.

and bring about stable conditions.

Initially, the majority of gold seekers, who endured untold hardships and found no gold, stated the amount of gold on the Fraser had been grossly exaggerated. Discouraged and penniless, they returned to San Francisco, denouncing the Fraser River as a humbug.

But the Fraser River was a genuine discovery, and in the next 20-odd years would produce more than ·twice as much gold as Spain had obtained from the Americas. Production figures for 1858 vary widely; the Canadian Minister of Mines reckoned the output was less than $500,000, but McDonald, referring to the records of bankers and express companies, thought it closer to $2,150,000. The yield increased in 1859 and 1860, and the total output for the three years was $6,000,000 or $7,000,000.

Late in the fall of 1860, as the Fraser River discoveries were beginning to fade, John Rose, Ben McDonald, 'Doc' Keithly and George Weaver made a discovery about 20 miles from Keithly Creek that would change the history of the Cariboo region of British Columbia. During the winter more than 400 miners were camped in the snow waiting for the spring thaw. When the snow melted, the early prospectors began their search for the precious metal. They were amply rewarded.

"At Antler Creek nuggets could be picked out of the soil by hand, and the rocker yielded 50 ounces in a few hours. Shovelfuls sometimes contained $50 each. Individuals were making $1,000 a day, and the output of sluice and flume claims was 60 ounces a day to the man. Much of the ground yielded $1,000 to the square foot."

Here, as in California, the best deposits were located in ancient stream beds, deep below the surface, which were reached by shafts and levels. The official yield for 1861 was $2,500,000, although official estimates range as high as $5,000,000.

In the late summer of 1862, a sailor named Billy Barker entered the Cariboo. Unable to find unstaked ground on Williams Creek, he decided to try below the canyon, much to everyone's amusement. But the laughter ceased when, at the 52-foot mark of his shaft, Barker struck pay dirt. And what pay dirt—his claim produced $600,000! The town of Barkerville, now a famous gold rush town tourist attraction, sprang up around his claim.

Following Barker's success, others also focused their attention on the area. Among them was John "Cariboo" Cameron, whose claim became the richest in the Cariboo. During 1863 it averaged from 120 to 336 ounces a day, with a lifetime output of nearly $1,000,000—at a time when gold was only worth $20 an ounce! Others were equally successful. "At Williams Creek several claims realized 100 ounces a day. One man obtained 387 ounces in a day and 409 ounces on the day following. At Barkerville the Diller Company washed out 52 ounces from a pan full of dirt.

"At Van Winkle Ned Campbell and associates took out 1,700 ounces in three days washing, and near there the Discovery Company, consisting of four men, took out 40 pounds in one day, and cleaned up at the end of the season with $250,000. At Lowhee Creek Richard Willoughby worked a claim on a blue slate bed rock within four feet of the surface, and obtained 84 ounces in one day and $1,000 in the week, while near him two brothers named Patterson took out $10,000 in five weeks, one day yielding 73 ounces, partly in nuggets weighing 10 ounces each."

From 1862-64 some 200,000 ounces were produced from the Cariboo. The Caledonia and Neversweat claims on Williams Creek yielded $750,000 and $120,000 respectively; Butcher's Bench ($125,000), Forest Rose ($480,000), and the Prairie Flower ($100,000), being but a few of the successful claims.

By 1865 the creeks were still yielding more than $3,000,000 annually. But shallow diggings were becoming exhausted, and by the late 1860s, the mining steadily declined. Yet, despite the vast quantities of gold produced in the Cariboo, it did not create worldwide attention, as the California and Australia gold rushes had done. The reason is thought to be because of the early denunciation of the Fraser River discoveries.

❀   ❀   ❀

ALASKA
Gold had been known to exist in Alaska as early as 1848. The first discovery is credited to Russian mining-engineer P.P. Doroskin.

The beginning of Circle City, Alaska, 1894.

Although his discovery in the gravels of the Kenai River generated considerable excitement, apparently no mining was carried out. In 1863 gold was discovered on Birch Creek by Rev. Robert McDonald, a Hudson's Bay Company employee. Although samples of coarse gold were sent to England, the Birch Creek deposits failed to generate much interest. Placer gold was discovered a third time in 1865-66. Like the two previous occasions, however, the Seward Peninsula deposits did not arouse much interest.

It was not until 1869, when gold deposits were located at Windham Bay and Sumdum Bay southeast of Juneau, that gold mining truly began. Alaska's first gold production came from this region, and during 1870-71 the total value extracted was reported to be worth $40,000.

In 1880 lode deposits were discovered at Juneau, and by 1883 it was the mining centre of the territory.

Then, in 1893, the Birch Creek deposits, which had originally been located 30 years earlier, were re-discovered by two Indians. When news of the find reached Fortymile, some 170 miles distant in Yukon territory, prospectors abandoned their meagre claims for the new prospects. "Throughout the summer of '94," wrote Lewis Green in *The Gold Hustler*, "these, and others, scoured the swampy flats and tributaries of Birch Creek. Along the banks, and on the sandbars, of the area's creeks, Mastodon, Deadwood and Greenhorn and others, they made not one, but many strikes; all of them richer than those at Fortymile, and sparking an even greater rush to the banks of Birch and its neighbours."

During the next few years an estimated $1,000,000 was produced from these diggings, and the community of Circle City grew into a sizeable town of log houses and buildings. This remained the most important mining camp in Alaska until 1896, when it was depopulated by the rich discoveries in the Klondike, an event that would dramatically alter the course of northern history.

❀ ❀ ❀

YUKON

The Yukon country was originally explored by the Hudson's Bay Company, which discovered gold as early as 1860. To protect the

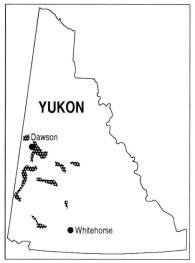

fur trade, however, the discoveries were kept secret. George Holt reportedly made discoveries along the Hootalingua River in 1878. However, although the territory embracing the White, Stewart, Pelly, Lewes and Hootalingua rivers produced $100,000 in 1885, the diggings were abandoned when coarse gold was discovered on the Fortymile River in 1886. "In 1887 and 1888, between 100 and 350 miners were at work on the Fortymile; production was about $100,000 in 1887, falling to $20,000 in 1888 because of the continued high water throughout the summer." Production increased slowly with the new discoveries on Sixty Mile, Miller, Glacier, Birch and Koyukuk rivers.

Meanwhile, a discovery was about to be made that would focus worldwide attention on the Klondike. Robert Henderson, who had prospected in many North American goldfields before making his way to Alaska and the Yukon, had halfheartedly prospected the Yukon River and its tributaries. None had proved promising until he struck pay dirt on a creek he christened Gold Bottom. In July 1896, while returning to his claim with supplies, Henderson met George Carmack, his Indian wife and his two Indian kinsmen, Tagish Charlie and Skookum Jim at the mouth of the Klondike River and informed them of his discovery. When Carmack asked permission to try the area, Henderson replied that

Discovery Claim on Rabbit Creek, Yukon, in the summer of 1897.
Carmack renamed it Bonanza Creek after he discovered gold.

"he" was welcome, but that he did not want any "damned Siwashes" staking claims on the creek. This was a brutal insult to Carmack's wife and kinsmen and, unknown to Henderson, was about to cost him an enormous treasure.

"Casually following Henderson's route up the Thron diuck," wrote T.W. Paterson in the *Ghost Towns Of The Yukon*, "Carmack and company came to a small tributary known as Rabbit Creek, and decided to try it out. Their efforts yielded them about 10 cents to a pan, which was not at all discouraging. But, before setting to work in earnest, they thought it best to visit Henderson's Gold Bottom."

What happened during that second meeting has never been entirely clear, although its almost certain Henderson protested Carmack's native relatives again. In any event, Carmack claimed he asked Henderson to join him on Rabbit Creek, while Henderson maintained that Carmack was to let him know if he found any rich diggings.

Whatever, upon their return to Rabbit Creek, Carmack soon discovered a ridge of bedrock which protruded above the soil. They found a thumb size nugget concealed in a crack, and produced four dollars from a single pan. Overjoyed, they immediately staked claims then, leaving Skookum Jim on guard, Carmack headed for Fortymile to record the discoveries. Although Carmack willingly informed everyone he met about the find, no attempt was ever made to notify Henderson. He remained isolated at his original claim on Gold Bottom, not learning about the rich strike being made just over the mountain until it was too late.

After the gold was exhibited at Fortymile, the news flashed throughout the territory. That same afternoon Fortymile lay deserted and abandoned, a virtual ghost town.

Joe LaDue, who is thought to have grubstaked Henderson, was one of the first on the scene. Primarily a businessman, LaDue located a town-site and named it Dawson City. In less than two years some lots on the main street sold for as much as $5,000 a frontage foot.

"Great effort was required to determine the real value of any location. Usually there was very little gold on the surface, except in cases like Carmack's discovery, where the bottom of the ancient stream bed happened to be exposed. Otherwise a shaft had to be sunk to bedrock where nature's concentration had placed the gold. But the ground was completely frozen. The miners built fires in their shafts to thaw out a foot or two of ground. Then, amid the smoke and fumes, they would dig the thawed-out earth and stockpile it, waiting for spring when they could run it through a sluice box at the creek. They managed to do some sampling by taking home a pan full after each fire and panning it out in a tub of unfrozen water in their cabin.

"During the winter of 1896-97 there was about 1,500 people in Dawson City. All early settlements along the Yukon, such as Forty Mile and Circle City, had been drained. Conditions in Dawson approached a famine. With all this going on, no real record of it reached the outside world."

But that situation was dramatically changed on July 15, 1897, when the *Excelsior* steamed into San Francisco Harbour with more than a ton of gold aboard. Two days later the steamship *Portland* docked in Seattle with an additional two tons—bringing one of the greatest and richest goldfields ever known to the attention of the civilized world. But this was only the beginning. During August, September and October, steamer after steamer brought back men laden with golden wealth, until $2,500,000 had been put into circulation. Within a year some $5,600,000 had been extracted from Eldorado and Bonanza creeks alone. The response to all this golden wealth was instantaneous, electrifying the world and touching off a global rush to the Klondike. Countless men, women and some children set out for the untold riches of the world's greatest gold rush. In *Klondike Fever*, Pierre Burton states that, "One hundred thousand persons; it is estimated, actually set out on the trail. Some 30,000 or 40,000 actually reached Dawson. Only about half that number bothered to look for gold, and of these only 4,000 found any. Of those a few hundred found gold in quantities large enough to call themselves rich. And of those fortunate men, only the merest handful managed to keep their

A panoramic view of the town-site of Gold Bottom at the junction of Gold Bottom and Hunker creeks, 1899.

wealth." Although the entire gold production of 1898 did not exceed $10,000,000, it is reported that prospectors seeking the precious metal had spent some $60,000,000! Obviously, there were many losers and only a few winners.

Yet, the richness of the Klondike deposits has seldom been equalled. Single claims produced $150,000 and $200,000 during the winter of 1896, and in the spring the owners declared they had worked only small corners of the mines. Single pans of dirt (two shovelfuls) yielded $800 to $1,000, and pans containing $300 to $500 were not uncommon.

The richest creeks of the bonanza districts included Bonanza, Eldorado, Victoria, Adams, McCormacks, Ready Bullion, Nugget Gulch, Bear, Baker and Chee-Chaw-Ka. Main Fork, Hunker and Gold Bottom creeks were in the Hunker district.

In the first decade, more than $100,000,000 in gold was mined around Dawson, and the Yukon thrived. It was made a separate territory in 1898, with Dawson, then a town of almost 30,000, serving as capital. Then, as large mining companies brought in complicated equipment to exploit the less accessible gold deposits, the placer miners began to leave. Gold valued at more than $22,000,000 was produced in the Klondike in 1900, the peak year. By 1910 most of the gold seekers had left the Yukon, and, by the 1930s, Dawson's population had dwindled to about 1,000 people.

After the heyday of the Klondike gold rush, placer production declined. Production was revived during the depression because the price of gold increased and unemployed men could still make a few dollars re-working old ground. Placer operations virtually ceased at the beginning of the Second World War. After 1950, improved machinery, which allowed greater yardages to be handled by fewer men, inspired a new revival. But by 1966 all large dredging operations in the Yukon had ceased. In recent years, activity in the Yukon, as well as other areas of historic rushes have enjoyed a significant increase, spurred by the high price of gold. And it would appear from all indications that this new gold rush will continue for many years to come.

An interesting sidelight to the Klondike gold rush concerns one Ed Schieffelin, who is best known as the discoverer of the original Tombstone claims. Although a wealthy man by 1882, Schieffelin was a dedicated prospector. To him, finding out where nature hid its golden wealth was like solving a challenging and perplexing mystery. Schieffelin spent a great deal of time studying maps of the Western Hemisphere in an effort to unravel the secret. He was already aware of the mineral wealth of Peru and Mexico, and he was familiar with the gold discoveries along the western states and British Columbia. By 1883 the first hints of gold in Alaska and the Yukon began to filter back, and, although the gold discoveries were still only vague rumours, the possibility was sufficient to lead Schieffelin to a fascinating theory. By pursuing his maps, he observed that the gold rushes in Peru, Mexico, the western United States and Canada lay in the same general line of mountains. He thus concluded "that a great mineral belt spanned the earth from Cape Horn, up through the mountainous ridge of South America, across Central America, and continuing along the continental divide northward to the Arctic lands where Alaska and Asia almost join.

"With the true prospector's disregard for financial security, he invested much of his profit from his Tombstone properties in a nearly-forgotten expedition to the frozen northlands. He built a little steamboat, outfitted a party, and in the spring of 1883 left the old Russian port of St. Michael on the Bering Sea to steam up the Yukon and penetrate the inaccessible hinterland. Somewhere, he felt sure, the belt of gold must cross this mighty river."

Although Schieffelin searched the Yukon River and the mouths of streams and creeks emptying into it for a thousand miles, his efforts proved fruitless until he reached a gorge known as the Lower Ramparts. Here he discovered sufficient traces of gold to conclude that he had found the beginning of the great mineral belt he had envisioned. But the gold lay in barren and forbidding territory, frozen-over eight months of the year. Schieffelin concluded that these obstacles were insurmountable, and that the gold could not be easily or profitably extracted. Dejected, Schieffelin returned to the states, married and settled down. He died in 1896 shortly after the great Klondike strike was made, but before news of it had reached the outside world.

This is an excellent example of how common sense, research and ingenuity can lead to great discoveries. Had Schieffelin displayed a little more confidence in man's ability and willingness to cope with the hostile northern environment, the Klondike gold rush would undoubtedly have been initiated much sooner and its history written differently.

PART II
METHODS OF PLACER MINING

I N this section I will discuss some of the more popular methods of placer mining. While a variety of methods will be examined, the reader must realize that not all of these methods will be appropriate or legal in every area. The laws pertaining to placer mining in Canada is different from those of the United States, and even the laws of one province or state may vary considerably from a neighbouring province or state. So potential placer miners are cautioned to learn the laws pertaining to the area where they will be working. Some of the devices used in recovering placer gold may cost hundreds, even thousands, of dollars. It is only common sense, therefore, to make sure a certain method is permitted before equipping yourself with the necessary materials.

THE GOLD PAN

L ONG synonymous with placer gold recovery, the gold pan is not actually a method of placer "mining" in the true sense. During the hectic gold rushes of California, British Columbia and the Yukon, it was possible for the individual miner to get a good return by simply panning the rich ground. But those days are long gone, and unless an ancient stream bed or undiscovered gold creek is found, panning for gold will never prove profitably worthwhile. The reason for this is obvious. Even an accomplished professional can only process about a cubic yard of material in a 10-hour day; the amateur considerably less. If that cubic yard only contains $10 worth of gold, your hard work will only return a paltry $1 an hour.

But the gold pan does perform some very important functions, and should therefore be a part of every prospector's equipment. First, it is the basic tool for testing the creek for gold. If panning reveals traces of gold, the prospector should then move upstream and test again. By this testing, the prospector will eventually find an area worth working by some other, more productive method that will permit more gold-bearing gravel to be processed in the same 10-hour day. A rocker, for example, will process three to four times as much material as a gold panner, while even a small sluice can process 10 times as much. It is up to the individual miner to determine the method best suited for him and the amount he is willing to pay to properly equip himself. Obviously, the more serious the miner, the more elaborate and complete his mining equipment.

The gold pan is also used to "clean-up" the concentrates recovered by other mining methods. It is also the favourite tool for the weekend prospector who is more interested in panning for

GOLD PANNING TECHNIQUES

(A) After filling the pan ¾ full of material, submerge it in water. Then knead it as you would bread dough, allowing the gold to sink to the bottom.

(B) Then remove any rocks or large pieces of material and discard.

(C) Give the pan several vigourous shakes back and forth and from side to side, or hit the pan sharply with the open hand. This will help move the heavier concentrates to the bottom.

(D) Hold the pan just below the water and tilted slightly from the body. Begin rotating the pan from side to side, allowing the lighter sand to be flushed over the lip.

(E) Level the pan occasionally, shake it back and forth, checking for nuggets. Then continue the panning procedure until nothing remains in your pan but black sand and concentrates. These should be stored for later processing.

fun and exercise than gold mining. However, if the miner is serious about recovering gold, he will only use the gold pan for testing, snipping or working very rich gold, if any of the latter can still be found.

Because I covered gold panning techniques in my earlier book *Gold Panner's Manual*, and because, as stated, gold panning is not a true "mining" method, I will not repeat myself here. However, if the reader has not yet learned to pan, I suggest you do so before going on to the more advanced methods about to be discussed.

WINNOWING &
THE PUDDLING BOX

O NE of the oldest—certainly one of the crudest—methods
of separating gold dust and small nuggets from gravel
was practiced by the original Spanish miners in America. This method called for nothing more than a blanket, two men,
the gold-bearing concentrates, and a slight wind. Once the concentrates had been thoroughly dried, the coarse gravel was separated with a screening box. This was a strong wooden frame with
a screened bottom that was placed in the centre of the blanket.
Dirt was piled into the screening box, which was then agitated as

Winnowing was one of the crudest methods of separating gold.

METHODS OF PLACER MINING
39

it was lifted. Only the smaller material would penetrate the screen and remain on the blanket.

Two prospectors then took hold of each end of the blanket and stretched it until only slightly slack, then the concentrates were flung into the air. The breeze, in theory, did the rest by blowing away the lighter dirt, allowing the heavier gold to fall back into the blanket. Although simple, winnowing was slow and inefficient.

Later, when the blanket became worn, it was burned, the ashes being panned to remove gold trapped in the weave.

❀   ❀   ❀

PUDDLING BOX

The puddling box is a wooden box about six-feet-square by 18 inches deep, arranged with plugs for discharging the contents. First, the box is filled with water and gold-bearing clay or cemented-gravel. The operator then stirs the material with a rake until it is broken up or dissolved. The muddy solution is then run off, leaving the concentrated material in the bottom of the box. These concentrates are then removed and washed in a gold pan in the usual manner, or in a rocker.

THE
ROCKER

A FTER the gold pan, the rocker is the simplest method of recovering placer gold. A rocker (see FIG. 1), sometimes called a cradle or dolly, is a simple contrivance consisting of a box about 42 inches long, 16 inches wide and 12 inches high, which is set on rockers (G), much like a baby's cradle. On the upper end is a removable tray or hopper (A), 18 to 20 inches square and four inches deep. The bottom of the tray consists of an iron plate perforated with half-inch holes (C). Beneath the hopper, and below the perforated plate, there is a light frame placed on an incline from front to back (D), upon which a canvas or burlap apron (E) is stretched. To operate the rocker, material is first thrown into the hopper. Then water is poured over the gravel with a dipper held in one hand, while the other hand rocks the cradle by means of the handle (H). The rocking movement must not be too violent, as you will defeat your purpose by washing the gold out along with the gravel. While a rocker can be operated by one person, it is more efficient when operated by two men. One man shovelling in the gravel and dumping the tray after the gravel has been washed, and the other rocking and bailing water.

The principle of the rocker is simple: the water washes the finer material through the bottom of the hopper and the gold is trapped in the apron below, or collects in the bottom of the rocker where two riffles are located (B). To save wear and tear, the floor of the rocker should be covered with galvanized sheet iron. The rocker is placed close to the bank of a stream or where water can be bailed or piped into the tray. The water flow should be as steady a stream as possible, since a sudden rush of water tends to wash away the fine gold. If the gold-bearing gravel is cemented or contains clay, a layer of mud may form on the burlap-covered

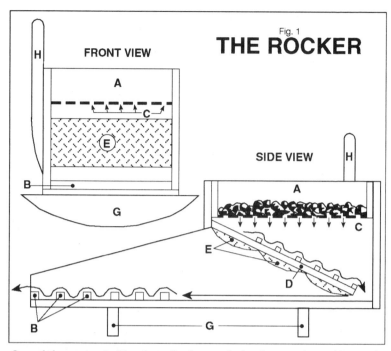

One of the most primitive, but effective, methods of recovering gold was the rocker, shown below at King Solomon mines, Yukon, in 1898.

apron, making gold recovery difficult. If this happens, the clay or cemented-gravel should first be stirred in the puddling box previously discussed. However, if the gold is mixed with the clay, the residue at the bottom of the puddling box will require separate panning.

Before discarding the lighter material and rocks left in the hopper, the waste material should first be examined, as nuggets over the half-inch size—you should be so lucky—cannot fall through the holes. It is also wise to pan the tailings occasionally to determine if gold is being lost.

Rockers were extensively used in both the United States and Canada before the introduction of the sluice. The main disadvantage of the rocker, much like the gold pan, is its lack of volume, as it is only possible to wash three or four yards of gravel a day. However, in its favour is the fact that the rocker requires little water with which to operate. In areas of limited water supply, the faithful rocker could be the only practical solution.

Constructing a rocker is relatively simple and inexpensive. A detailed plan was published by the Department of Mines and Petroleum Resources in *Bulletin No. 21* (see Fig. 2). These plans are for a small type rocker which is used in recovering coarse gold. A larger rocker is more efficient for recovering fine gold because of the length of its sluice, but the small type can be more easily transported from one location to another. Since large rockers should be disassembled for easier transporting, they are constructed as "knock-down" units, held together with nuts and bolts instead of nails.

The following information, regarding construction of a rocker and relating to Fig. 2, was also published in *Bulletin No. 21*.

Fig. 2 (1) is the side view of a rocker showing the two-by-four-inch side-braces nailed to the side-boards of the box. One of these is extended and tapered for a handle. Each side of the box and sluice can be cut out of one piece of one-by-12-inch lumber, 42-inches or more in length. The bottom of the box, (2), can be made of one piece of board 16 inches wide and 42 inches long. If not procurable, two pieces planed so that they fit tightly together can be used. It is safer to cover the bottom of the sluice with canvas, galvanized iron, or tin to prevent leakage, and, in the latter case, assist the flow of sand and gravel. The tray, which is built of one-by-6-inch lumber, 17 inches long, with screening or a punched galvanized plate nailed to the bottom or held in place with a one-by one-half-inch wooden strap, is set upon two two by two-inch supports nailed to the side of the box at an angle sufficiently great so that when the entire rocker is set at the proper gradient it will tilt slightly forward. Make the outside measure-

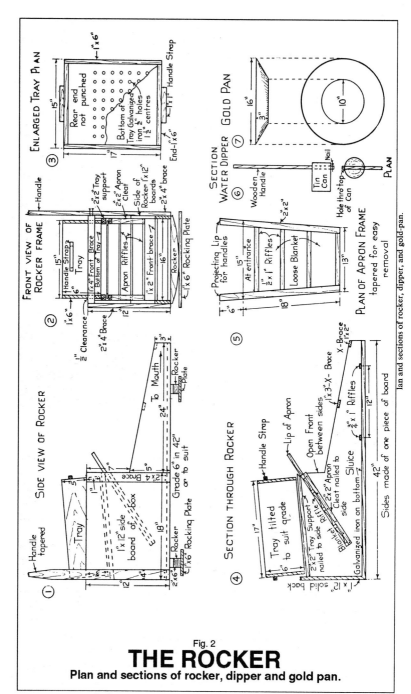

Fig. 2

THE ROCKER

Plan and sections of rocker, dipper and gold pan.

ments of the tray small enough so that it can be removed easily. Two pieces of wood nailed on the ends of the tray will be useful for handles. Be sure that the boards used for the rocker are free from knotholes, otherwise gold will be lost.

At the bottom of (1) and (2), two "rockers," made by two-by-six-inch or two-by-four-inch lumber, the width of the sluice, and bevelled from the centre outwards, are nailed to the box sufficiently far apart so that it can be rocked to and fro easily. Underneath, two rocking-plates or flat stones are laid to keep the rocker in place. In some rockers a steel pin or large spike is inserted in the centre, which fits into a loose socket bored in the plate. In this way the box is kept from slipping down-grade.

In (2) the front view of the rocker-frame is not drawn to scale, but to show the construction of the different supports, etc., clearly. For instance, the bottom of the tray is, in reality, hidden by the one-by-four-inch brace.

Details of a rocker, showing perforated screen and metal riffles.

In (3) an enlarged drawing of a tray is shown. The rear end of the tray can be punched if the apron is built nearly the full length of the box. If not, it is better as planned so that the gold will fall upon the blanket riffles before being washed down the sluice. The slight down-gradient given to the tray will generally be sufficient to move the gold over the punched holes.

In (4) the position of the tray before being tilted to obtain a suitable grade is shown. Also, the approximate position of the blanket riffle, which must be set on a steep enough gradient so that there will be as little packing of gravel on the riffles as possible. Two or three sluice-riffles are generally sufficient, but more can be added if the tailings are found to contain gold. Two cross-braces are necessary to keep the top of the sluice from warping.

In (5) the plan of the apron shows the projecting lips of the frame which are useful for pulling out the apron before the clean-up. The tapered measurements can be regulated to suit the size of

Washing gold with a rocker on Spruce Creek in B.C.'s Atlin district.

the box. If the frame is not tapered, it may stick owing to fine gravel packing along the sides. The loose blanket can either be tacked on or held in place with a narrow strip of wood; in some operations the blanket is used alone without wooden cross-pieces. The sand packed behind the riffles should be stirred occasionally so that the gold can sink.

Drawing (6) shows a long-handled dipper which can be constructed by punching a hole through the top of a can and driving a nail as shown to keep the can from slipping.

THE TOM, SLUICE & GRIZZLY

THE Tom succeeded the rocker. It consists of two separate troughs, or boxes, placed one above the other, which rest on logs or stones, and have an incline of about one inch per foot (FIG. 3). A sheet iron plate, or riddle (C), perforated with half-inch holes, covers the lower end of the first trough (B), which is bevelled on the lower side. This first trough, known as Tom proper, varies from 15 to 20 inches in width at the upper end, and is 30 inches wide at the lower end. It is usually eight inches deep. The second trough (D), is actually a box sluice with common riffles (E). The overall length of the Tom was about 12 feet. The Long-Tom was usually around 14 feet. Again, to save wear, the bottom of the Tom is lined with one-eighth-inch sheet iron.

To operate the Tom, a stream of water enters Tom proper through the spout (A), just above the point where the gold-bearing gravel is being introduced. The gravel is usually shovelled in by one man, while the second constantly stirs it about with a square-mouthed shovel or fork with blunt tines, pitching out the heavy boulders and throwing back undecomposed lumps of clay and cemented-gravel against the current. The washed gravel, on striking the riddle, is sorted, the fine dirt and muddy water passing through the holes. The gold and other concentrates are then trapped against the riffles in the second trough. When the riffles are full, the concentrates are removed and panned in the usual way. Larger rocks which are stopped by the riddle must be tossed aside frequently to prevent clogging.

THE SLUICE

By far the most important arrangement for mining placer gold is the sluice. Introduced shortly after the Tom, it is now almost universally employed for the collection of gold from placer deposits.

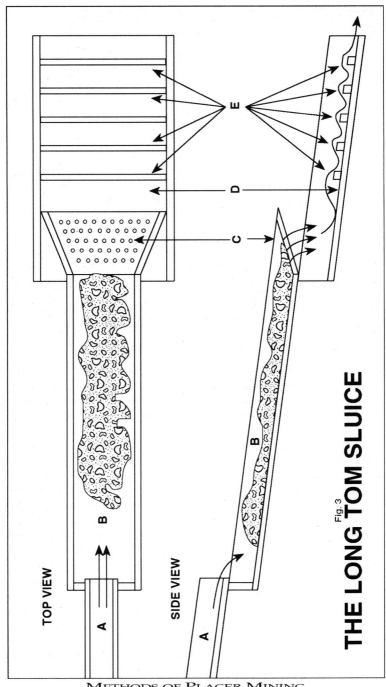

TOP VIEW

SIDE VIEW

THE LONG TOM SLUICE

Fig. 3

A long-tom sluice with one man bailing the water and the other shovelling gravel. Coarse coco-matting covers the bottom of this particular sluice box.

There are two types of sluices: box sluices—raised above the ground, and ground sluices—sunk below the surface. Box sluices, or board sluices, are long open-ended wooden troughs, or a series of troughs, that are joined together to form sluices that vary in length from 50 to several hundred feet. Each sluice-section is 12 to 14 feet long, usually 16 to 18 inches wide, and eight to 12 inches deep. Sluices are rarely narrower than 12 inches, and can exceed six feet. They are usually constructed of one-and-one-half-inch planks, with the lower end tapered so that the narrow end of one box telescopes into the broad end of the next throughout the whole series. Beyond this, no nailing or fixing is required. The sluice rests on trestles, and usually has a uniform descent throughout the entire length. This incline or descent, commonly known as its "grade," is usually 10 to 18 Inches per 12 feet in length. The grade is regulated by the position and length of the apparatus and the nature of the dirt to be washed. As a general rule, however, a fall of less than 10 inches, or more than 18 inches on the length of a 12-foot box, is not suitable for the ordinary sluice.

A false bottom is used to catch the gold and save wear and tear on the floor proper. These false bottoms—preferably of metal

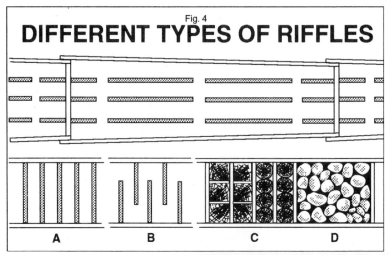

DIFFERENT TYPES OF RIFFLES

Fig. 4

A   B   C   D

due to the wear and tear—are frequently made of longitudinal riffle bars six feet long, three to seven inches wide, and two to four inches thick. Two sets are required for each length of sluice, and these are kept in position by cross-wedges, at a distance of one to two inches apart. They are never nailed, as they must be removed during the clean-up.

❀   ❀   ❀

HOW TO BUILD AND OPERATE A SLUICE

To operate a sluice, gold-bearing gravel is shovelled into the channel. A stream of water flows continuously through this gravel, washing and decomposing it. The lighter material is then carried away, while the gold and other heavy concentrates are trapped in the spaces formed between riffles, the gold always sinking through the lighter material to the bottom.

Sluice box riffles, showing canvas covered by coarse expanded metal screen and a section of Hungarian riffles below.

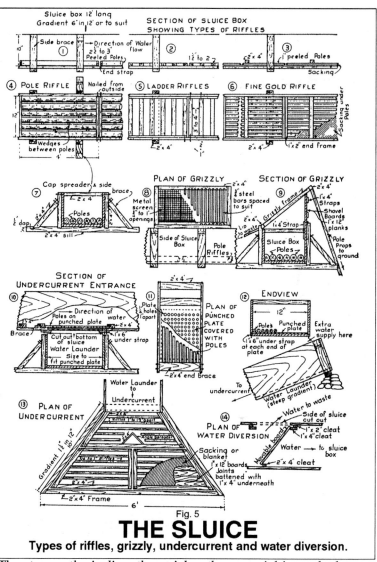

Fig. 5

THE SLUICE
Types of riffles, grizzly, undercurrent and water diversion.

The steeper the incline, the quicker the material is washed away by the force of the water; the tougher the dirt, the steeper must be the grade, as tough clay naturally does not break up as quickly in a slow current as in a rapid one. In short sluices, therefore, the incline should be relatively slight, as there is more danger of losing fine gold in a short sluice than in a longer one. Some sluices begin with a steep descent, then almost level off towards the end. This effectively washes and disintegrates the toughest material,

HOW TO BUILD A SLUICE BOX

Fig. 6

STEP 1

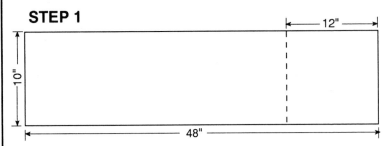

STEP 1
Saw a 12" piece off the four foot 1"x10" board. This leaves the three feet necessary for the bottom of the main box, and provides a one-foot section for the holding tray.

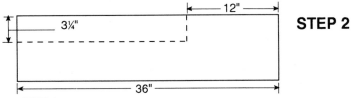

STEP 2

STEP 2
Using two pieces of three-foot long 1"x8" boards, measure 12" along the top and 3¼" down the edge and cut out as per the dotted line.

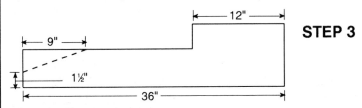

STEP 3

STEP 3
Measure 9" from the narrow end, and 1½" from the bottom. Draw the angle and cut according to the diagram.

STEP 4
Nail the two sides to the outer edge of the bottom section (the three-foot 1"x10" from Step 1. Then nail an eleven inch 2"x4" to the bottom rear of the unit.

STEP 5
The holding tray is then installed by positioning the one-foot piece of 1"x10" (cut off in Step 1) and nailing it in place. A piece of eleven-inch 1"x4" board is then cut and nailed across the opening on the back to enclose the holding tray on three sides.

STEPS 4 & 5
Assemble the pieces as shown below.

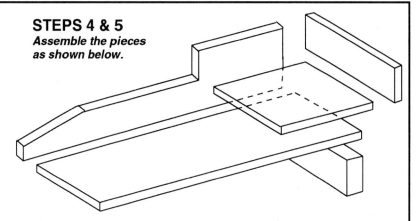

STEP 6
The riffle tray is constructed with two twenty-two inch 1"x1" pieces. These are positioned in the main box, taking care to allow a slight space for water expansion. The distance is then measured to the outside edges of the parallel strips and two pieces of molding (¾"x¼") are cut and nailed at either end. Then remove the riffle frame and install the remaining riffles between the two at the ends. They should be placed about 1⅛" apart.

STEP 7
Indoor-outdoor carpet is then fitted into the bottom of the sluice box. This will trap the fine gold that would otherwise escape.

STEP 8
The riffle tray is then repositioned in the sluice box over the carpet. Four ¼" diameter holes are then drilled through the 1"x1" pieces on the riffle tray, through the carpeting and the bottom of the sluice box. You then install with bolts which are tightened down with wing nuts.

Your completed sluice box is now ready for use.

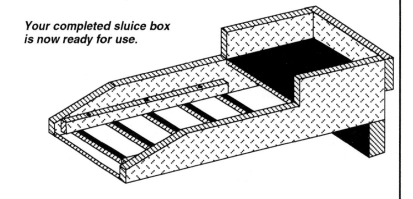

then allows the gold to sink as the water slows near the end. It should be remembered while sluicing that, generally, about 80 percent of the gold is caught in the first 200 feet. Sluice boxes employ the use of various types of riffles. (see FIG. 4 and FIG. 5). The plank riffles illustrated in the top drawing of FIG. 4 can be effectively replaced by logs. These are often readily available at the site, resulting in less material having to be transported to your workings. When rocks are plentiful in the gravel you are washing, it is wise to place a small bar across the end of each trough to prevent riffles and the bottom of the sluice from being run bare.

Other types of riffles used in sluices include: FIG. 4 (A) common riffles, (B) zigzag riffles, (C) block riffles, and (D) stone riffles.

Common riffles are merely pieces of wood extending from one side of the trough to the other. These form obstructions which impede and collect the gold. Common riffles vary in size, depending upon whether they are used in a rocker, Tom, sluice or dredge.

Zigzag riffles are similar to common riffles, except that they do not extend all the way across the trough. Instead, the alternate ends are open, forcing the water to swing around them like the curves of a stream. Some pioneer prospectors swore by this type of riffle, claiming that it was the most efficient for recovering the finest values.

Block riffles are used when there are a lot of pebbles and boulders in the gravel being washed. These rocks tend to wear down normal riffles too rapidly, so block riffles, which last much longer, are substituted. These are wooden blocks, eight to 13 inches deep, cut across the grain, and placed upright in rows across the sluice. Each row is separated by a space of one to two inches, and kept in place by riffle strips. Square block riffles are the best for saving gold, although sections of wood are frequently used just as they are sawn from the round log. Whichever is used, these riffles offer a further advantage of trapping the finest gold within the grain. The miners of old, upon the block riffles wearing out, would burn them and pan the ashes.

Stone riffles are used to the best advantage where heavy, strong cemented-gravel is being washed. The rocks are naturally of an irregular shape and size, and are set in the sluice with a slight tilt downstream. Rock riffles are cheap and extremely durable, but have the disadvantage of being more awkward and costly to handle, requiring a longer time to clean up and repave the sluice. They also require steeper grades and more water.

In FIG. 5 reprinted from *Bulletin No. 21*, (1), (2), and (3) sections of peeled longitudinal poles, cross or ladder, and fine-gold riffles are shown through the side of the sluice box. Constructed from poles two or three inches thick and between three and four feet long, the pole riffles are laid lengthwise and about half an inch apart at the head of the sluice. The ends of the poles are nailed to two cross-pieces of lumber or split poles (4), the diameter of the riffles and the width of the box. Wooden wedges are driven between the poles to keep them apart at the centre. The cross-pieces are nailed to the side of the box from the outside, leaving the head of the nail projecting so that it can be withdrawn easily. Nails driven inside the box will be pounded by the gravel so that they cannot be removed for clean-up.

The cross or ladder riffle (S) is generally placed next to the poles. The cross pieces may be made of squared poles or lumber, one-by-two-inches, and placed about four inches apart. The ends are fastened to two longitudinal one- by two-inch straps four feet long built like a ladder so that the riffle section can be easily removed.

A fine-gold riffle (6) may be placed at the end of the sluice box. First a blanket or double sack is laid at the bottom of the box.

A prospector washing gravel with a grizzly at Edmonton in 1890.

On top of this is a small peeled-pole framed riffle with wedges is placed. This is designed to protect the burlap blanket from pounding by the gravel and to retain any gold that has settled to the bottom. The two-by-four-inch cross riffles on top of the poles are generally placed about three feet apart. This causes the necessary swirl in the water to thoroughly wet the fine gold so that it may sink.

❦   ❦   ❦

THE UNDERCURRENT, OR GRIZZLY

All of the previous methods discussed were tried-and-true, and more than one early miner earned a king-sized fortune through their use. But, as the gold streams were systematically picked clean of their treasures, the miners found that it was more and more difficult to recover the finer gold values which remained. As a result, the more elaborate "undercurrent" came into being. This refinement of the traditional sluice was considerably more effective than its predecessor, being able to recover values as fine as 1/2000 of a cent.

The undercurrent employs an ordinary main sluice box, but in addition four or five identical sluice boxes are positioned alongside and parallel to the main sluice, with a common feed trough at right angles. Each of these auxiliary sluices has a gradient of one-half inch to one-and-a-half inches per foot, and usually employs the favoured zigzag riffles. The gravel to be processed is shovelled into the auxiliary sluice, from which it works its way into the feedbox, thence through a further screen, or slatted frame called a grizzly, into the main sluice for final processing. The result, as noted, is an extremely high rate of recovery.

A grizzly, FIG. 5 (8) and (9) is a set of parallel steel bars or rails, or even poles, set in a two-by-four-inch rack and spaced according to the size of gravel which is to be kept out of the sluice box. The grizzly is set over the head of the sluice at an angle of about 45 degrees so that the oversize material will toll off easily. The upper side is held in place at the required height by two pole props nailed to the frame of the grizzly and stuck in the ground. If a wheelbarrow is used, a wooden hopper can be built over the grizzly and the water-supply line raised to the height of the hopper so that the boulders will be washed away before dumping. If the boulders are covered with mud, it is advisable to hand-wash or puddle the larger boulders and put the remainder through the sluice box as the mud may contain gold.

As the main sluice is situated at the greatest angle, the resulting mix of gravel and water from the other sluices passes through at a fairly rapid rate, thereby concentrating the coarser nature of the gravel. The undercurrent, or grizzly method, proved to be not

This grizzly was being used on the North Saskatchewan River, Alberta, in 1898.

only the most efficient of the early day methods, but offered an additional bonus in that the main sluice yielded as much gold as it would have yielded when operating by itself. The additional values which were recovered by the adjoining undercurrent sluices and grizzly came as icing on the cake.

There are many types of undercurrent sluice boxes which are successful. A box of this type is designed to save fine gold that otherwise would be washed away in the ordinary sluice by the disturbance caused by coarse rolling gravel. There are two important rules for the beginner to remember: (1) That fine gold must be wet before it will sink; (2) that when it has sunk it must be undisturbed, consequently coarse gravel must be kept out of the box in which the fine gold has fallen to the bottom.

In Fig. 5 (10) a section through the side of the sluice box shows details of the position of the intake for an undercurrent. First of all, the bottom of the sluice box is cut out of the required length and width of the screen, consisting of a plate punched with one-quarter inch holes one inch apart. Underneath the box and projecting far enough into the hole cut so that the plate will rest safely upon them are nailed two one-by-six-inch pieces of lumber, the width of the sluice box. After placing the plate in position, a riffle, made of framed, small peeled-poles spread about a quarter of an inch apart with wedges, is placed lengthwise over the plate and nailed in place from the outside of the sluice box. Congestion often occurs in any type of undercurrent entrance, especially if the gravel contains much black sand, so care must be taken to keep the punched plate clear. The two-by-four-inch riffle shown on the upper side of the plate opening can be set at the proper distance, according to the volume of water used, to give the final swirl.

In FIG. 5 (12) the undercurrent water-launder is shown in section. An extra amount of water is often necessary to drive the black sand on to the undercurrent. The gradient of the launder delivering to the undercurrent can be adjusted with rocks as shown.

In FIG. 5 (13) one of many types of undercurrents is shown. The idea, generally, is to spread the product received from the intake over as large an area as possible to give it a chance to settle more readily than it would in an ordinary sluice box. The angle-spreader, made of two pieces of two-by-four-inch, set on edge near the top of the undercurrent, assists in accomplishing this, as well as keeping the product away from the junction of the two separate frames below.

In FIG. 5 (13), the bottom of the undercurrent is made of one-by-12-inch boards, planed, if possible, and fitted closely together. Underneath, one-by-four-inch strips are nailed to the joints to prevent as much leakage as possible. On top of the boards a blanket or sacking of canvas or burlap is laid. On top of this are set three quarter-to-one-inch riffles about one-and-a-half inches apart. Be sure that the hairs of the blanket or sacking do not project over the riffle. Instead of wooden riffles, metal webbing, if procurable, can be used to advantage. The dimensions given are arbitrary and any convenient size can be used. Plenty of water and a gradient of at least one-and-one-half inches in 12 inches is necessary. All riffles must be constructed as separate trays so that they can be easily removed for clean-up.

Drawing (14) shows one plan of water diversion used when cleaning up the sluices. The side of the sluice box is cut out so that the piece removed will fit across the box and divert the flow of water through the opening.

If water is scarce and insufficient to roll the gravel quickly down the sluice box, a space of a few feet is left between the riffles and is covered, if possible, with a metal plate or flattened tin can. This will accelerate the flow of water.

In cleaning up, only a small flow of water is allowed to run through the boxes. The riffles are removed successively from the top end and washed in the box itself. By working over the concentrates in the bottom of the box with long-handled wooden paddles it is possible to wash off additional light material. As the concentrates "stream down" the sluice, the gold begins to show at the head end. This should be picked up by means of a whiskbroom and small metal scoop. Finally, the concentrates should be shovelled from the box and the remaining finer gold panned or rocked from them.

If the gold is very fine, it may be separated from the concentrates by amalgamation.[1] About five times as much mercury should be used as there is gold. If the gold is coated with oil or grease and will not amalgamate, it should be cleaned with soda or lye. Rusty, tarnished gold will not amalgamate, unless it is cleaned by abrasion in a barrel amalgamator. An amalgam-barrel should be used if the operation yields fairly large amounts of sluice-box concentrates containing fine gold or gold that is difficult to amalgamate.

When using mercury, care should be taken to prevent loss as it is expensive and escapes readily.

The proper method for separating the gold from the mercury after amalgamation is by means of metal retort.[2] This can be obtained from any chemical supply company. The mercury, volatilized by gentle heat, will pass up through the tube in the sealed cap of the retort and down the tube into a bucket of water. Care must be taken to prevent the escape of the mercury fumes, which are poisonous, and for the same reason retorting should be done in the open or at least in a well ventilated room.

The early miners used a halved and partly hollowed potato into which the amalgam was placed. The halves were then bound tightly with wire, and the potato was put into a fire or oven. The heated mercury would evaporate into the potato, leaving pure gold.

---

1 For readers not familiar with amalgamation, it is covered in detail in *Gold Panner's Manual*.
2. See the chapter "How to Recover Fine Gold" in *Gold Panner's Manual*.

THE MODERN SLUICE BOX

WHILE the general shape of the sluice has not changed much over the years, dramatic improvements have been made in the materials with which it is constructed and the efficiency of its riffle design. This permits a three-foot modern sluice to out-perform the 12-foot type used by early prospectors. Early sluices were telescoped one into another and could be up to a quarter-mile in length, while the modern sluice is only a few feet long. Early sluices could require 15 to 20 men shovelling material at one time, while the modern sluice, is easily operated by one man. The riffles of the early sluices, as already mentioned, were conveniently constructed of any handy material, from poles to rocks, while the favoured modern riffle is the "Lazy L" or Hungarian Riffle. This design creates an inner "vortex," sucking the heavier gold and material to the base of the riffle. And while the early sluices recovered 80 percent of the gold passing through it in the first 200 feet, the modern sluice retains 90 percent of the gold in the first six inches!

While it is relatively easy to build your own sluice box of wood (see FIG. 6), there are a number of drawbacks against wooden construction. First, even a "water-proofed" wooden sluice, because of gravel running through it, saturates with water relatively quickly. Second, while a dry wooden sluice is heavy enough to carry, a water-logged sluice is considerably heavier. Third, water will eventually rot a wooden sluice, so that it may be useless after a few uses.

Modern sluice boxes, on the other hand, are constructed of tough plastic or aluminum alloys. While incorporating all the features of the old, they are extremely lightweight and portable; a blessing to those who have to hike into the back country.

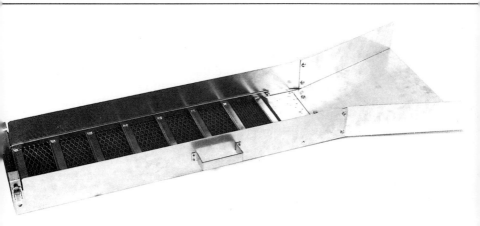

The Keene Model A52 Hand Sluice. This unit was designed with a wide flare for faster separation and made longer for greater capacity. It weighs 11 pounds.

Four lightweight aluminum models are manufactured by Keene Engineering of California, the largest manufacturer of gold-mining equipment in the world. Weighing from five to 11 pounds respectively, these mini sluices can process up to 10 times as much gravel as the gold panner. Two similar models, weighing 11 and 25 pounds, are manufactured by Placer Equipment Manufacturing of Arizona.

To operate, you merely select a spot along the stream with easy access, proper incline, and a minimum water depth of four inches. You then place the sluice directly in the current so that the box is filled with water up to the top of the trough. The proper "grade" should be one inch for each running foot of the sluice. To test the current's ability to carry gravel through the sluice, scoop up a handful of material and drop it into the upper end of the trough. If the current washes the lighter gravel through the sluice quickly, you have found a good location.

Once the sluice box is set up, a gold pan for cleaning-up the concentrates, a small pointed shovel or garden trowel, a bucket and a large spoon are about the only other tools required. The prospector simply shovels or pours the gold-bearing gravel into the top end of the sluice, in carefully regulated amounts. Do not dump an excessive amount of gravel into the sluice box all at once. This will overload the riffles and permit the sand on the top to wash right through the sluice, carrying gold with it. The gravel must be fed at a pace that allows you to see the "crest" of each riffle bar at all times. You must also continually check the riffle section for large waste rocks which are not being washed through the sluice. These must be removed by hand because if the riffles get clogged, the water will wash out the concentrate in the immediate area, and with it, the gold.

When your riffles have accumulated black sands or concentrates in amounts extending more than half-way downward to the next lower bar, it is time to perform a clean-up. This should be performed once or twice a day unless you are in an area of extremely heavy black sand or gold. To do this, simply remove the sluice from the stream, being careful to keep it level, and carry it to the bank. Here the sluice's riffle section is removed and rinsed off in the large bucket. The matting which lines the bottom of the sluice is then removed, rolled up and placed in the bucket where it is thoroughly rinsed. Finally, the sluice itself should be rinsed into the bucket. The resulting material is later panned and amalgamated to recover the gold. The sluice is then re-assembled and placed back in the stream.

THE
DREDGE

WHEN most people think of dredges, they visualize the large house-sized floating monsters that were used in the Yukon and other gold-bearing areas, or the huge machines used in deepening or widening river channels. But, except for a few rotting derelicts, the monstrous gold dredges of the past are gone forever, replaced with modern pieces of equipment that can be easily handled by one or two men.

The modern gold dredge performs somewhat like an underwater vacuum cleaner, sucking up gold-laden gravel from the creek bottom and washing it over riffles which trap the gold.

Those who recover the greatest amount of placer gold today will unquestionably be armed with a dredge. This type of prospecting has many advantages for the gold miner. First, because the units are generally light-weight, portable, and completely self-contained, they can be easily back-packed into the less accessible areas. Second, depending on the size selected, a miner can reach and clean out bedrock cracks in deep pools that would be inaccessible by any other means. Third, dredges can process a massive amount of material each hour; even an entire bar of gold-laden gravel is not beyond the capacity of the correct-size dredge.

Basically, there are two main types of dredges; surface and underwater. Both operate on the Venturi principle to create the vacuum necessary for dredging. A gas-powered engine drives a high-velocity centrifugal pump which sucks in water through a screened intake valve. A length of high-pressure hose connected from the output of the pump then carries the water to an eduction system, which creates tremendous suction. The two types of eduction systems commonly employed are the suction nozzle (FIG. 7), and the powerjet.

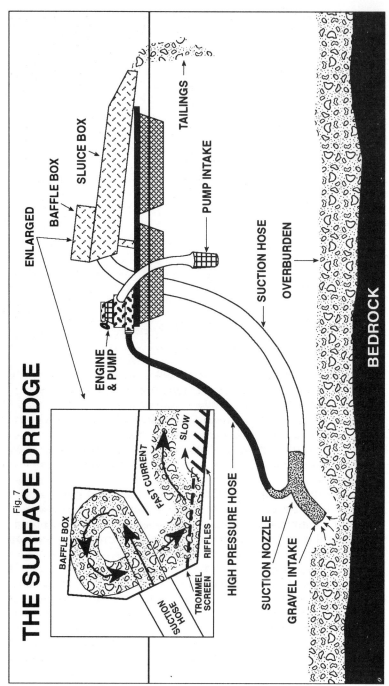

THE SURFACE DREDGE

Fig. 7

ENLARGED

BAFFLE BOX

SLUICE BOX

TAILINGS

PUMP INTAKE

ENGINE & PUMP

SUCTION HOSE

OVERBURDEN

BEDROCK

BAFFLE BOX

FAST CURRENT

SLOW

SUCTION HOSE

TROMMEL SCREEN

RIFFLES

HIGH PRESSURE HOSE

SUCTION NOZZLE

GRAVEL INTAKE

Keene Model 2003 two-inch backpack dredge. It is powered by a heavy duty 2-hp, 2 cycle engine. The inflatable pontoons may be rolled up into a small package for storage and transportation. It weighs only 45 pounds.

The suction nozzle is a curved piece of tubing attached to the gravel intake at the end of the suction hose. When water from the pump is forced through an orifice in the curve, it creates a high-velocity water stream up the suction hose. This high-velocity water displaces the water in the suction hose and creates a suction at the mouth of the gravel intake. When this intake is pushed into the gravel, this suction draws it in and forces it up the suction hose.

While the principles of the suction nozzle and powerjet are very similar, the powerjet is the most advanced and efficient means of creating suction in a modern gold dredge. The main difference is that in the suction nozzle water is forced through an orifice near the bottom of the suction hose, while in the powerjet, the water enters the suction hose through an orifice much nearer the sluice box. These location differences mean, basically, that in a suction nozzle the water-gravel mixture is pushed through the suction-hose, while in the powerjet the mixture is pulled through.

A surface dredge (FIG. 7) consists of a sluice box, engine and pump floating on the surface of the stream on inner tubes or moulded plastic floats. The surface dredge is the most popular and can use either a suction nozzle or powerjet eduction system.

THE UNDERWATER DREDGE

Fig. 8

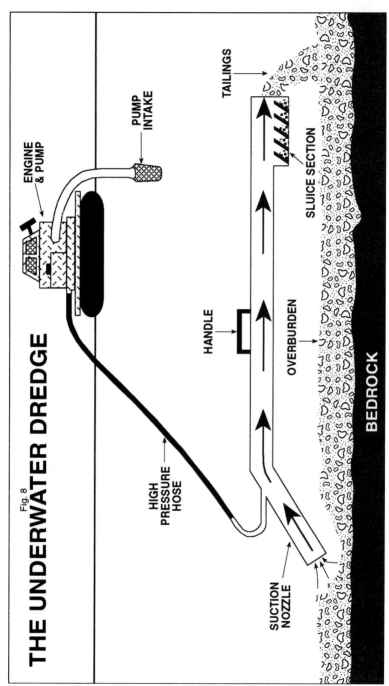

ENGINE & PUMP

PUMP INTAKE

HIGH PRESSURE HOSE

HANDLE

SLUICE SECTION

TAILINGS

OVERBURDEN

BEDROCK

SUCTION NOZZLE

In each case the water-gravel mixture is sucked from the bottom of a river to a sluice box mounted on the dredging unit itself.

Before the gravel reaches the riffle section of the sluice box, it must first enter a hopper or baffle box (FIG. 7, inset). In most cases the water-gravel mixture enters the baffle box at a 45 degree angle. When it strikes the wall of the baffle box the direction is reversed in a churning action for classification (separation) over a trommel screen located about one inch above the floor of the sluice. As the mixture is agitated in the turbulent action of the baffle box, any cemented material is quickly broken up. When the mixture hits the trommel screen the smaller material passes through the half-inch holes for a more selective separation in an area of less water velocity. The riffles then separate the gold. Since the surface dredge has a large riffle section, it has the distinct advantage of being able to recover fine gold. The sluice box will function the same in either type of eduction system, although the flow of water is far greater with the powerjet because of its proximity to the sluice.

Another advantage of the surface dredge is that the tailing pile of processed material can be discarded further from your work area. With the underwater dredge, the spot you are working has a tendency to slowly refill with tailings.

An underwater dredge is completely submerged except for its gas-powered engine which floats on the surface (FIG. 8). It basically consists of a suction nozzle welded to a light-weight metal tube with a riffle section at the end. The water is forced into an orifice at the suction nozzle under pressure exactly the same as the surface dredge. But here the similarities end. Instead of the water-gravel mixture being carried to the surface, it is carried through the long metal tube and over the riffle section. Because the material does not need to be carried to the surface, the underwater dredge can move large amounts of overburden. A 6 inch underwater dredge with a 5-hp engine will move the same material it would take a 15-20-hp surface dredge to do.

Both Keene Engineering and Placer Equipment manufacture dredges in a variety of sizes and models. Both companies manufacture the following sizes: 2 inch—45 pounds; 2½ inch—70-95 pounds; 3 inch—120-130 pounds; 4 inch—175-230 pounds, 5 inch —280-350 pounds and 6 inch—425-1,300 pounds. Keene also manufactures two 8 inch dredges, one weighing 2,000 pounds and the other 2,500. Needless to say, the larger the dredge the more material it will process in a given time. The 2 inch dredge will process about two yards of gravel per hour, while the large 8 inch dredge will sift through up to 30 yards of material per hour. But you must remember, as the capacity of the dredge increases, so does the size, the weight, and also the cost. And, while larger dredges can process considerably more material, they require two men to operate efficiently. The 2 inch dredge is the next logical progression from the sluice. It weighs about 45 pounds and is mainly used for "sniping." With the mini-dredge you can easily vacuum crevices and bedrock cracks, recovering material too deep to reach with a pick. It is not designed for moving sand bars or gravel banks.

Dredges of six inches or larger are professional pieces of equipment designed for moving gravel banks and other large volumes for gold recovery. When you get into the larger dredges, it then becomes practical to move a large mass of material to get at the richer gravel on the bottom. If a known stream runs $3 or $4 worth of gold per cubic yard of gravel, for example, the very bottom material will be 10 to 20 times richer. But you may have to move 20 cubic yards of overburden to get at a single cubic yard of bottom. The larger the dredge, the faster you accelerate the process.

Somewhere between the 2 inch and 8 inch size is the correct dredge for you. But how do you determine which one? Before purchasing your first dredge you might consider the following. First, 80 percent of the people who purchase gold dredging equipment intend to use it strictly as a form of recreation. Second, a small dredge can process approximately two cubic yards of gravel per hour—about the same as the professional gold panner can process in 10 hours. Third, the miners who are doing well with the gold dredge are the ones that put a light-weight model on their back and hike into some of the more inaccessible, high-producing areas that have not been worked for years. Fourth, a 45-pound, 2 inch modern dredge can easily perform the same task it took a 300-pound unit just over a decade ago. When these facts are considered, it would seem that the best way for the novice to start is with a small, inexpensive unit. Then, as his interest increases, it is much easier to determine the optimum size unit for his purpose.

The user of a small dredge should confine himself to areas where there is no more than four or five inches of overburden. Remember, the size of a dredge is determined by the inside diameter of the suction hose, not the intake opening, as most people believe. To prevent rocks that are too large from entering the suction hose where they might get clogged, the intake of the nozzle is usually 25 percent smaller.

When the cubic yard capacity of a dredge is mentioned, it refers to optimum conditions and uniform "sized" gravel. If you purchase a load of gravel, each "piece" is generally about the

(Above) The Keene 1993 series of 4 inch portable dredges are more powerful and efficient than its previous models.

(Below) The new 6 inch Gold King dredge from Placer Equipment Mfg. The sluice box is 24"x6"x8", and the unit is powered by twin 8 hp Briggs & Stratton engines.

The Keene Triple Sluice Dredges. Available in 4, 6 and 8 inch models, this design is capable or recovering values never before obtainable in a single or double sluice recovery system. This machine separates, sizes and classifies material before it enters the recovery system. Gold bearing material less than a half-inch in size is diverted into the side sluices for gentle washing and concentration.

same relative size. In a stream, however, this is seldom the case. The two cubic yard capacity of the 2 inch dredge, for example, is based on "sized" gravel no larger than one inch. Rocks or boulders larger than this must first be removed by hand. Since this will naturally cut down productivity, the actual capacity will be somewhat less. Similarly, the capacity of a 6 inch dredge would be based on "sized" gravel no larger than four inches.

As the size of the dredge increases, so does the engine. As the horse-power goes up, it also becomes practical to add a compressor so that you have underwater breathing. You can now work in water three to six feet deep, putting yourself out of reach of the smaller dredges. The majority of gold dredging, incidently, is done in 10 feet of water or less.

Dredging should always be conducted by at least two men. One diver sucks the dirt from the river bottom while the second watches the sluice, signalling the diver when it is full. The second man can also come to the rescue of the diver in case of accident.

Curiously enough, unlike the old expression that two is company and three is a crowd, when it comes to dredging the most efficient system consists of a three-man team. With three men on the job, two act as divers and the third as the surface operator. The first diver handles the mouth of the dredge, sucking up gravel as he moves it along the stream bed, while the second diver, equipped with a crevising gun, shovel, crowbar and such, proceeds along the bottom in front of the dredger to remove large boulders from his path, and crevising by hand between those too big to be moved. He also breaks up clay-gravel conglomerate which otherwise would not be picked up by the dredge. On the surface, the third man controls the dredge, advances the float as required, and signals his partners below to rest while he empties the riffles, as necessary, into a five-gallon can for further refining by hand.

Unfortunately, underwater dredging invites greater risks on the part of the prospectors involved, and stringent safety precautions should be observed at all times. Under no circumstances should SCUBA divers attempt to dredge in swift-flowing waters, or when the water temperature dips below 45 degrees (Fahrenheit). Even when equipped with the insulated diving suits, the frigid water will soon have its effect; and chilling can lead to more serious problems. Even in summer, most mountain-fed streams are cold and a limit of 10-15 minutes should be set on all dives.

When underwater, divers must be aware of large boulders which might topple over on them; they should play it safe and ignore all rocks larger than a one-gallon bucket. Above all, divers should never work without a surface man. Even a minor accident underwater can mean death. And, of course, minor injuries can prove to be fatal as, chances are, it's a long way to the highway and to the hospital.

THE DRY WASHER OR ELECTROSTATIC CONCENTRATOR

O NE of the newest, best, and most intriguing methods of "dry" washing is the electrostatic concentrator. This device employs air and gravity, rather than the water of the traditional sluice box. It also eliminates the inconsistency of the pounding action used to help transport material in the "bellows" type of dry washer.

The electrostatic concentrator owes its development to that hoary method of dry washing discussed at the beginning of this section—winnowing. Although the original means of winnowing, as described, utilized a blanket and a breeze, a more refined version used a wool blanket which improved the rate of recovery by the fact that the blanket, as it was continuously tossed up and down, created a build-up of static electricity. This helped retain the finer gold and other precious metal particles.

With progress, the wool blanket has developed into the modern-day wonder which consists of a gas-powered high-static air fan to force air into a plastic trough or container, where it gains an electric charge from the plastic. From there the air moves under pressure through a special artificial fabric where the charge increases even further.

The gold-bearing material is

The Keene DW2, hand crank model dry washer weighs only 32 pounds and is very portable.

Manufactured by Placer Equipment, the Gold King "Dust Devil" is a gas powered dry concentrator with many technical advances. The unit above, which weighs 85 pounds, will process two yards of material per hour.

shovelled into the compressor through a trommel screen. Made of aluminum and high-strength plated-steel, the trommel screen separates the larger material much like the grizzly of old, and allows only gravel less than a half-inch in diameter to enter the concentrator. The segregated material then travels down the plastic tray over the synthetic cloth and a set of hinged riffles.

The secret of this type of concentrator is the fact that gold, although not magnetic, has an affinity for an electrostatic charge which causes the gold to actually cling to the cloth before passing into the riffles. As the refined gravel works its way through the riffles, it is bombarded by other particles of material, thereby causing a natural concentration of the heavier material, just as occurs in an ordinary sluice box with running water. The static electricity constitutes the initial recovery process, which is followed by the sorting of material as it continues over the riffles in the secondary, and final, recovery stage.

The Keene Electrostatic Concentrator Model 151 has been carefully designed. The portions which handle the electric charge are thoroughly insulated to prevent loss of charge by ground leakage. The frame is of high-strength plated tubular steel, giving maximum rigidity consistent with the lightest possible weight.

The frame is designed so that the overflow tailings drop directly on both front legs. This provides a stability usually found in machines many times its weight. The entire unit disassembles quickly into a compact package for carrying. Assembly takes only a few minutes.

The trommel screen is constructed of durable aluminum and high-strength plated steel—again, a combination designed for light-weight strength. Power for the system is generated by a high-static air pump connected to a 3½-hp engine. Both engine and pump are aluminum, and the mounting system permits heat from the engine to transfer directly to the air pump. This preheats the air, making it possible to process damp sand that cannot be put through other units. This heat transfer procedure also has the added advantage of creating cooler engine operating temperatures, giving it extended life under continuous duty in nearly any environmental condition. A long flexible hose connects the engine-pump unit with the concentrator, so that the power pack may be placed on the up-wind side, thereby avoiding the clogging and abrasive effects of dust from the concentrator.

A simple turnbuckle arrangement provides for complete adjustment of the angle of the concentrator tray to accommodate all types of material.

Normally, clean-up of the riffles is done every hour, at which time the operator can gauge the richness of the deposit he is working by examining the gold and black sand concentrates which have collected. Using a small brush, he clears the surface gravel from the top of the first or second riffle, when, satisfied as to the productivity of that site, he empties the tray of concentrates into a gold pan for final working.

According to its manufacturer, this model will recover gold up to 200 mesh and can process in excess of two tons of gravel per hour. Its total weight is only 75 pounds. But for those who consider this unit too large, Keene has also developed a mini-dry washer. Weighing only 32 pounds, the Model DW2 can be folded so small it is easily transported in a back-pack. It will set-up in seconds and, when operating at optimum efficiency, can process about one ton of material per hour.

By pulling a cord, air is forced from a bellows through a cloth, setting up an electrostatic charge to attract the gold. The DW2 operates 50 percent on the electrostatic principle and 50 percent on the specific gravity principle, giving complete gold separation and recovery. The bellows is constructed of a high-quality vinyl-reinforced nylon which should give you years of trouble free service.

THE HYDRAULIC CONCENTRATOR

U NLIKE the electrostatic concentrator, the complete hydraulic version weighs in at from 80-110 pounds. It is designed to process placer gravel above the river level, such as during low water in the summer, or along the banks of a river which has been lowered over the ages. Extremely portable, they are rated as being capable of handling about two tons of gravel per hour, for the smaller units, up to 12 tons for the largest units, and they will recover gold, platinum and other precious minerals as fine as 200 mesh.

More than anything else, the hydraulic concentrator looks like a glorified mini-sluice, and works in much the same way. When in operation, water is pumped to the concentrator through a meter valve which enables the operator to control the water flow as desired. A plastic container dissipates the force of the water from the pump so that it spills over the "classifier" screen without too much turbulence which could wash away the fine particles of gold.

The sluice unit is positioned on a three-inch grade. When the gravel is shovelled into the hopper it is instantly saturated with water and segregated according to size, the larger material washing over the screen as the finer material falls into the sluice for final separation.

The Keene Hydraulic Concentrator is constructed of high-quality aluminum and high-impact plastic. It is designed to be quickly set up once the prospector has arrived at his location. Disassembling the unit is equally quick and simple.

To clean-up, the riffles—which are a single unit—are removed. The expanded metal screen is taken out and the water-proof matting carefully removed. Most of the concentrates, which

(Above) Powered by a 3-hp Briggs and Stratton engine, the Keene Model 173 includes a pump unit and 25 feet of one-inch pressure hose so that it can be set up away from the source of water. A suction hose, foot valve and strainer complete the unit. It weighs 82 pounds.
(Below) The Gold King High Banker-Dredge, manufactured by Placer Equipment, features a 5-hp Briggs & Stratton engine.

have collected in the matting, are emptied into a gold pan for final recovery.

❀ ❀ ❀

HYDRAULIC CONCENTRATOR AND DREDGE COMBINATION

For the prospector, especially the novice, who cannot decide what system to purchase, this item should be of interest. Both Keene Engineering and Placer Equipment each manufacture fully-equipped dredges that double as a light-weight hydraulic concentrators. The Keene 165-3 offers the prospector the versatility and advantages of two different systems in one compact unit, enabling him to clean-up bedrock cracks or go after the higher bench and residual placers.

Keene Model 173-25.

The Keene 173-25 is the model 173 dry washer which is converted to a dredge with the addition of 10 feet of a 2½-inch dredge suction nozzle and 12½-feet of additional high pressure hose. The unit includes a centrifugal pump, sluice box and four hoses. Because the engine is shock-mounted, it will provide stable performance wherever it is set up. The pump draws water through a 1½-inch diameter hose with a foot valve and strainer. The foot valve, which must be submerged at all times, should be in relatively clean water. The sluice box consists of a hopper with a plastic header, classifier screen with half-inch holes, long and short support legs for the suitable angle, one-inch meter valve, suction hose mounting bracket and a Hungarian riffle tray.

When being used as a dredge, the 2½-inch suction hose is attached to a metal coupler which in turn fits into a mounting bracket above the screen of the sluice. At the end of the suction hose is a suction nozzle creating the suction discussed with the surface dredge.

The material being dredged is deposited in the hopper where the heavier gold-bearing gravel falls through the classifier screen onto the riffle section of the sluice. When it is time for clean-up, let the suction nozzle draw in clean water for a few minutes. This will wash out both the hopper and the riffle section, leaving only the black sand concentrates. Loosen the wing nut at the bottom of the sluice box, lift up the one-piece aluminum riffle section, and remove the waterproof matting. The concentrates should be care-

Keene's Hydraulic Concentrator in action.

fully rinsed in a bucket for later gold panning.

The hydraulic concentrator uses the same engine and pump, and the same basic set-up as the dredge system, The difference, of course, is that instead of vacuuming up gravel from a creek bottom, you are shovelling gravel into the hopper. The hydraulic system then washes and classifies the gravel. Instead of removing the pressure hose from the pump to couple to the suction hose, the concentrator uses a 25-foot one-inch diameter hose which couples to the pump and runs the meter valve at the back of the hopper. In a typical set-up, the operator will put the engine and pump near a source of clean water, and carry the unit to the location of the gold. Since the sluice unit only weighs 17 pounds, this is a relatively easy task. As with the hydraulic concentrator previously discussed, the flow of water entering the sluice is controlled by the operator via the meter valve. It is also dissipated in a plastic container at the rear end of the hopper. This provides a steady flow of water to wash the material being deposited in the hopper. The recovery process from this point is the same as in the dredge system.

This combination gold-recovery system weighs only 66 pounds but can pump 125 gallons a minute.

TROMMEL CONCENTRATOR

The "Gold Claimer," manufactured by E.F. Domine Company of Minnesota, is a compact, self-container gold placer concentrator. With a capacity of two to eight cubic yards per hour, the unit is ideal for testing gold values or for a small production operation. The Gold Claimer has its own water supply and recirculating system, is eight feet long, less than three feet wide and is less than five feet high. It has a total weight of 600 pounds, runs off a 3½ hp Tecumseh air-cooled engine, and is mounted on a 13 gauge steel base frame that makes it fully portable, with brackets for trailer mounting.

The Gold Claimer does not require an extra pump or hoses, but when coupled with a 12-inch by 12-foot conveyor and a small loader, the operation can process up to eight tons per hour. By working gravels assaying one-tenth of an ounce of gold per cubic yard, an operator could bring in three to eight ounces of gold per 10-hour working day.

Efficiency is a key ingredient to the Gold Claimer. A spray bar in the trommel delivers up to 3,950 gallons of water per hour to scrub the gravels. It also disintegrates the clay to shake loose micron gold and scalps off over-sized particles. The sluice, which oscillates at 150 pulses per minute, keeps the sand bed open. Rif-

The Gold Claimer, shown here on trailer mounted with lights and 4.80 wheels.

fles and astroturf on the sluice can quickly be removed by releasing four toggles, providing for fast and efficient down time. In addition, the water reclamation unit also acts as a gold trap.

The power train on the Gold Claimer is built to be simple and easy to maintain. In the event of breakdowns, replacement parts are easily available. The Gold Claimer has a 100-gallon water capacity, which is enough for testing, but for high production, a tank or lagoon is needed.

DRIFT MINING

THROUGHOUT this book you will have encountered numerous references to "drifting" or drift mining. This method is not new. It was frequently employed by the early prospectors in areas where the gold was deeply buried by overburden. It often was, and sometimes still is, the only means of mining gold in some locations. Many of the placer deposits of the Cariboo, for example, were mined by this method.

Drifting presupposes the concentration of gold-bearing gravel in a well-defined stratum or channel. When the existence of such a pay channel is determined, it is mined by means of a tunnel run in such a way as to remove all gold-bearing material. The channel may be on bedrock covered by deep overburden, or may lead into a hillside. While both would be mined by drifting, each would be tackled differently.

If the channel is buried in the bottom of a valley, a shaft must first be sunk down to bedrock. From here drifts are driven along the bedrock following the paystreak. The gravel is raised to the surface by means of a large bucket. Billy Barker's famous claim on Williams Creek was mined in this manner. After sinking his shaft 55 feet Barker encountered ground paying $5 per pan. Bedrock was at 80 feet, however, and when this depth was reached, the paystreak was far richer.

If the channel is located on a bench in the hillside and rises as it enters the hill, the tunnel is run along its bed, following the bedrock; otherwise, the tunnel is driven through the "rim-rock," in such a position that the lowest point of the deposit will be above it. Gravel is removed from the tunnel either by wheelbarrow or ore car. The Neversweat Claim was drifted in this manner.

The following information on how to properly timber a drift

or tunnel is reprinted from *Notes On Placer Mining, Bulletin No. 21.*

"In starting to drive through unconsolidated material, a cut of sufficient size should be made so that the face will stand a few feet higher than the height of the working. If the ground holds up well, the cut may be completed before timbering. If the ground runs, timbering will have to proceed with the excavation. In running ground, a set may consist of two posts (uprights), cap, and sill

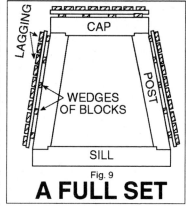

Fig. 9

A FULL SET

(top and bottom cross-pieces). (See FIG. 9.) The posts rest on the sill and the cap rests on the posts. The cap and sill are notched so that the posts will not slip under pressure from the sides. When drifting on bedrock, the posts are usually set directly on the bedrock, and mud sills are not needed. Bridges or strips of plank are placed outside of and parallel to the posts and caps, from which they are separated by small blocks, leaving spaces between the posts and bridges through which lagging can be driven. Lagging is made of sawn or hewn planks; small poles are less satisfactory.

"The lagging is placed outside the posts on the first set and outside of the bridges on the second set. The lagging from the second to third sets, and so on, is driven through the spaces between the posts and bridges of the second set slightly outward so that it will come outside of the bridges on the next set. A similar procedure is followed for the lagging driven over the caps. The lagging is driven forward as the ground is excavated. When the lagging has been driven forward about half of a set-length, a false set is placed to guide and support it. When the full set-length has been driven, the set is put in place, the lagging driven forward over the bridges, and the false set removed, allowing the lagging to press against the bridges of the completed set. The sets are usually placed four feet centre to centre and lagging five to six feet in length is used. Collar-braces should be put between sets.

"Shaft-timbering for preliminary work is carried out in much the same manner as timbering a drift in soft ground. The heavy timbers serve the same purpose as the posts, cap, and sill of a full set, but are called wall and end plates. The wall plates of the collar set should be long enough to project beyond the ends of the shaft opening so that they can be placed on a solid foundation. The second set is ordinarily suspended from the first set by hang-

The Sheepshead claim, Williams Creek, B.C., c1867,
showing shaft dug down to bedrock.

ing-bolts, and so on for each succeeding set. The prospector, however, seldom uses this method and simply wedges his sets into place. When the ground stands up fairly well, he can complete a set before lagging it up, but in the case of soft ground he drives his lagging in the same manner as in driving a drift through ground that will not stand without support.

"A simple headframe and windlass can be erected on the collar-plates and a bucket used to raise the muck. A ladder must be built in every shaft.

"If the shaft is inclined, skids must be placed for the bucket to slide on. Small peeled poles nailed to the wall-plates on the lower side of the shaft will serve the purpose."

Since water is the greatest problem in drift mining, the tunnel should be excavated at a sufficient grade to permit good drainage. A simple way to accomplish this is to dig a ditch along one side of the drift at a depth sufficient to allow the water to flow out. Where full sets are used (FIG. 9), the ditch will naturally pass beneath the sills.

The gravel in the drift is generally excavated by hand-pick, then removed by wheelbarrow or ore car. If a wheelbarrow is used, a wooden runway should first be laid down. This is easily constructed of split timbers or planks. When using an ore car with flanged wheels, wooden rails made of planks, surfaced with strap-iron for protection, will prove satisfactory.

Large fortunes were mined by drifting in the past, and with any luck, it could prove to be a worthwhile method even today.

THE SPIRAL
GOLD PANNER

PLACER mining can be a great outdoor hobby for the entire family. Its fun, exciting, and it might even put a few bucks in your pocket as well?

To do that, regardless of what type of gold recovery equipment you are using, or plan to use, the bottom line is the final recovery of values. In the past, this has always been accomplished by the old fashioned gold pan. But with today's modern design and engineering methods, the backbreaking labour of hours spent hunched over the gold pan in the icy waters of a stream are a thing of the past. Now, while you forge ahead with your sluice box, dredge, high banker, or dry washer, the spiral gold panner can be automatically performing the final separation of gold from your heavy concentrates. Or, if you prefer to do all your clean-up in the evening, you can relax around the campfire while your spiral panner contentedly works away at the backbreaking labour.

A spiral gold panner is a battery powered wheel with built-in spiral riffles and a recirculating water supply. Fine concentrates (all the rocks, pebbles, and larger gravels should be screened out) are introduced into the bottom of the spiral wheel where the material is picked up by the rotating riffles. A fine spray of water across the riffles (aided by a deliberately induced mechanical vibration) washes the lighter materials from the riffles and back into the bottom of the bowl, while the gold and heavier concentrates continue to spiral their way up and through a hole in the centre of the bowl where they are retained in the catch-cup.

The heart of any automatic gold panning machine is the spiral wheel itself. Camel Mining Products, of Arizona, manufacture five different models with bowl diameters ranging from 13½ to 38 inches. The Little Camel, Camel Deluxe and Spartan Camel, with bowls of 13½, 16½ and 18 inches respectively, each contain seven spiral leads on the wheel. The Pro Camel 24 and 36 models each have 15 spirals that lead into a five spiral second stage. The smallest unit can pan 100 pounds of concentrate per hour, while the largest unit can process 600 pounds. The recovery rate is about 95 per cent.

Gold Screw Inc., of California, manufactures an 18-inch unit that has four spirals. The more leads a unit has, the faster it can process material. This is simple mathematics. If a spiral panner had just a single lead, once the gold was picked up by a spiral, individual particles would have to travel about 28 feet before they could drop through the bowl's centre into the catch-cup. Now, if you add a second lead, individual gold particles would only have to travel 14 feet, thus cutting the recovery time in half.

The Gold Screw automatic panner is the only automatic spiral panning machine on the market to induce one of the primary principles of gold recovery—vibration—into their unit. On the back of the cast pulley you will find the heads of four bolts protruding. As the wheel revolves, these bolt heads press against a roller and induce a 100 TPM (Thump Per Minute) vibration. This "thump" closely resembles the action the old prospectors used to help settle the gold in their pans. The continual shaking of the pan and the constant bumping action of the side of the pan against the opposite palm helped the old sourdoughs quickly and efficiently settle the gold and heavy concentrates. The thumping action of the Gold Screw panner does precisely the same thing. The amount of vibration is easily controlled by simply adding a washer beneath the head of each of the four bolts if more vibration is desired, or by removing the cam wheel to stop the vibration completely.

Most of the automatic panners are well designed and efficient to operate and many units may be assembled or disassembled in a matter of minutes for transportation or storage.

BIBLIOGRAPHY

Garnet Basque, *Gold Panner's Manual,* 12th Printing. Langley, Sunfire Publications, 1993.

Pierre Burton, *Klondike Fever.* New York, Alfred A. Knopf.

Colliery Engineer Company *Placer Mining—A Hand-book for Klondike,* 1897.

Charles H. Dunning. *Gold—From Caveman to Cosmonaut.* New York, Vantage Press, 1970.

Bob Grant. *Product Report: The Gold Screw Vibrating Gold Panner.*

Lewis Green. *The Gold Hustlers.* Anchorage, Alaska Northwest Publishing Co., 1977.

Ernest Ingersoll. *Gold Fields of the Klondike.* Langley, Sunfire Publications (Mr. Paperback imprint), 1981 reprint of 1897 book.

Jerry Keene. *Gold in a Campground.* Northridge, Calif., Keene Industries.

A.H. Koschmann and M.H. Bergendahl. *Principal Gold-Producing Districts of the United States.* Washington, Geological Survey Professional Paper 610, 1968.

Robin May. *The Gold Rushes.* London, William Luscombe Publishers Ltd.

T.W. Paterson. *Ghost Towns of the Yukon.* Langley, Stagecoach Publishing Co. Ltd., 1977.

Matt Thornton. *Dredging for Gold.* Northridge, Calif., Keene Industries, 1977.

Klondike—The Chicago Book for Gold Seekers. Monarch Book Co., 1897.

Notes on Placer Mining in British Columbia (Bulletin No. 21). Department of Mines and Petroleum Resources.

LIST OF GOLD MINING EQUIPMENT MANUFACTURERS

KEENE ENGINEERING
9330 Corbin Ave.
Northridge, Calif. 91324
Manufacturer of gold mining equipment.

ROCKY MOUNTAIN DETECTORS LTD.
Box 5366, Sta. A
Calgary, Alberta T2H 1X8
Canadian distributor for Keene Engineering.

PLACER EQUIPMENT MANUFACTURER INC.
427 N. First St. (Miller Road)
Buckeye, Ariz. 85326
Manufacturer of gold mining equipment.

CAMEL MINING PRODUCTS
P.O. Box 3179,
Quartzsite, Ariz. 85346
Manufacturer of spiral gold panning equipment.

GOLD SCREW INC.
20144 Bassett St.
Canoga Park, Calif. 91306
Manufacturer of spiral gold panning and sluicing equipment.

PLACER RECOVERY SYSTEMS INC.
Ste. 140 7901 Flying Cloud Dr.
Eden Prairie, Minn. 55344
Distributor for E.F. Domine "Gold Claimer" and rotary riffles.